Visualising Place, Memory and the Imagined

This book probes into how communities and social groups construct their understanding of the world through real and imagined experiences of place.

The book seeks to connect the dots of the factual and the imaginary that form affective networks of identities, which help shape local memory and sense of self and community, as well as a sense of the past. It exploits the concept of make-believe spaces – in the environment, storytelling and mnemonic narratives – as a social framework that aligns and informs the everyday memory worlds of communities. Drawing upon fieldwork in cultural heritage, community archaeology, social history and conflict history and anthropology, this text offers a methodological framework within which social groups may position and enact the multiple senses of place and senses of the past inhabited and performed in different cultural contexts.

This book serves to illustrate a useful visualisation methodology which can be used in participatory fieldwork and thus will be of interest to heritage specialists, ethnographers and cultural geographers and oral history practitioners who will particularly find the methodology cheap, easy to replicate and enjoyable for community-based projects.

Sarah De Nardi is a lecturer in heritage and tourism at Western Sydney University and the co-editor of the *Journal of Community Archaeology and Heritage*. Her first monograph, *The Poetics of Conflict Experience: Materiality and Embodiment in Second World War Italy* (2016), revived landscape perception in Second World War Italy through hands-on oral histories with veterans.

Critical Studies in Heritage, Emotion and Affect
In Memory of Professor Steve Watson (1958–2016)

Series Editors
Divya P. Tolia-Kelly
Durham University
Emma Waterton
Western Sydney University

This book series, edited by Divya P. Tolia-Kelly and Emma Waterton, is dedicated to Professor Steve Watson. Steve was a pioneer in heritage studies and was inspirational in both our personal academic trajectories. We, as three editors of the series, started this journey together, but alas we lost his magnificent scholarship and valued counsel too soon.

The series brings together a variety of new approaches to heritage as a significant affective cultural experience. Collectively, the volumes in the series provide orientation and a voice for scholars who are making distinctive progress in a field that draws from a range of disciplines, including geography, history, cultural studies, archaeology, heritage studies, public history, tourism studies, sociology and anthropology – as evidenced in the disciplinary origins of contributors to current heritage debates. The series publishes a mix of speculative and research-informed monographs and edited collections that will shape the agenda for heritage research and debate. The series engages with the concept and practice of heritage as co-constituted through emotion and affect. The series privileges the cultural politics of emotion and affect as key categories of heritage experience. These are the registers through which the authors in the series engage with theory, methods and innovations in scholarship in the sphere of heritage studies.

Affective Geographies of Transformation, Exploration and Adventure
Edited by Hayley Saul and Emma Waterton

Museum Franchising in the Age of Cross-Border Heritage
Beyond Boundaries
Sarina Wakefield

Visualising Place, Memory and the Imagined
Sarah De Nardi

For more information about this series, please visit: www.routledge.com/Critical-Studies-in-Heritage-Emotion-and-Affect/book-series/CSHEA

Visualising Place, Memory and the Imagined

Sarah De Nardi

Routledge
Taylor & Francis Group

LONDON AND NEW YORK

First published 2020
by Routledge
2 Park Square, Milton Park, Abingdon, Oxon OX14 4RN

and by Routledge
605 Third Avenue, New York, NY 10017

First issued in paperback 2021

Routledge is an imprint of the Taylor & Francis Group, an informa business

Publisher's Note
The publisher has gone to great lengths to ensure the quality of this reprint
but points out that some imperfections in the original copies may be apparent.

British Library Cataloguing-in-Publication Data
A catalogue record for this book is available from the British Library

Library of Congress Cataloging-in-Publication Data
A catalog record for this book has been requested

ISBN 13: 978-1-03-208641-5 (pbk)
ISBN 13: 978-1-138-05227-7 (hbk)

Typeset in Times New Roman
by Apex CoVantage, LLC

Contents

Acknowledgements

I am extremely thankful to all my co-researchers, the community collaborators and allies without whom my life's work and this volume in particular would have been impossible. From the friends and colleagues at the Gruppo Archeologico del Cenedese Archaeoclub (Chapter 3) to the residents of Kibblesworth and Ryhope (Chapter 6), these remarkable individuals and their communities have been the lifeblood of my explorations of the emplaced nature of memory.

I am obviously indebted to the Engagement and Participation team at Beamish Museum, with their insightful work and deeply entrenched participatory ethics. In particular, I would like to thank Lisa Peacock and Geraldine Straker, without whom I would not have met the community co-researchers involved in the Remaking Beamish project. Working with these professionals gives me hope about what museums can achieve in the community. Many thanks to Sally Dixon in her capacity as assistant director of partnerships and communications and to Beamish assistant director Helen Barker and her team for a fruitful and hopefully ongoing collaboration. Warm thanks to Emma Sayer and Dan Hudachek of the Collections Department for showing me around the stores and talking to me about what the museum is collecting for the ongoing Remaking Beamish project.[1] Co-authoring a section of Chapter 7 with Geraldine Straker has been a pleasure: thank you for agreeing to write with me.

At Durham University, thanks go to my former Re:Heritage project partners Mike Crang and Nicky Gregson of the Department of Geography, with whom I have shared hours of stimulating conversation and imaginative exchange of ideas. Professor Divya P. Tolia-Kelly at Sussex University has been one of the best mentors anyone could ask for. A further thanks goes out to the Swedish project colleagues at the University of Gothenburg working in several departments, namely the School of Global Studies, the Centre for Consumer Science and the Department of Conservation. Project members Anna Bohlin, Staffan Appelgren, Anneli Palmsköld, Ingrid Martins Holmberg and Niklas Hansson have been a pleasure to work with. Thank you for welcoming me to your wonderful University of Gothenburg facilities, and thanks to the wonderful Anneli for guiding me to Swedish and Danish open-air museums and in particular to *Kulturen* in Lund (Kulturhistoriska föreningen för södra Sverige). Our trip was a real help in gauging how different countries approach social histories from below.

Over the English Channel, I am hugely thankful to my official and unofficial co-researchers, gatekeepers and esteemed colleagues at the Istituto per lo Studio della Resistenza del Vittoriese (henceforth ISREV): chiefly, my scholarly partners Pier Paolo Brescacin and Vittorino Pianca were instrumental to the design and implementation of the Second World War fieldwork and participatory research in Vittorio Veneto (chapter 4). I am currently planning further and more exciting work with you all.

Thank you! to Kate Cameron-Daum for reading early drafts of the manuscript and giving me precious and encouraging feedback. And warm thanks to my friend and one-time supervisor Ruth Whitehouse for helping me actualise the heritage field maps that appeared in my doctoral dissertation and for encouraging me to make use of community-based knowledge and perception in my thesis at a time when avocational collaborations were still frowned upon.

Note

1 The £20 million Remaking Beamish project is the biggest in the museum's history and will feature a 1950s town, a 1950s farm and expansion of the 1820s area, including a inn where visitors can stay overnight. The 1950s town will include a cinema being moved from Ryhope in Sunderland, homes, shops, a community centre, a cafe, a fish and chip shop and a bowling green. Aged miners' homes will provide a pioneering centre for older people, including those living with dementia. The expansion of the 1820s landscape will include the inn, a quilter's cottage and other examples of early industry.

Thanks to the money raised by National Lottery players, the Remaking Beamish project has been awarded £10.9 million by The National Lottery Heritage Fund.

Introduction

This book explores imaginative and multivocal ways of visualising senses of place, community-held values and memories. A series of field-based collaborative mapping practices illustrate and articulate the workings and potential merit of such methods (and many variations thereof). This book and the visualisations therein conceptualise place through the lens of affectual layers produced and shared by the expectations, imaginaries and ambitions of many. The actual mapping process is part of a wider framework of grassroots world-building and local place-making with communities, of which the mapping experiments and visualisations seek to capture ethnographic detail at the intersection of place, heritage, history and identity. Encountering *through* academic discourse and storytelling the workings of communities who imagine, remember and tell their own stories, we can perhaps hope to glean how social groups imagine and remember together.

Other leitmotivs of the present volume observe and critique affectual interplay among heritage, memory and place and the role of heritage co-production as imaginative practice. The implementation of the concept of co-production, initially articulated by Elizabeth Olstrom in the field of development studies (1996), found growing success in the humanities: more in the following. Indeed, the idea of collaboration as heritage-making process is central to this book's project. I view heritage is a social experiment, created and driven by community input, something that people 'feel' and 'do' as part of their everyday lives in places. The development of heritage is the development of connections. Further, heritage is what Watson and Waterton define as a 'process': "linked to memory, identity, politics, place, dissonance and performance" (2011: 4). Our connections to the past, contend Watson and Waterton, are "tangible, and [. . .] have a materiality upon which they depend that makes them objects of heritage" (2011: 2). Tangible or otherwise, heritage connections may be more or less strong or tenuous depending on our own personal investments, our own agendas and our own affectual stakes in these knowledges (see Orange and Laviolette, 2010; Waterton and Watson, 2015a). The intangible materiality of certain heritage practices may even replace physical presence (Orange and Laviolette, 2010) without compromising the bonds between people, place and the past.

I am most intrigued by the ways in which we, as listeners and scholars, capture the mood, the feeling, the here-and-now of participation and involvement. We are present at the moment in which the person or people we are talking with remember, relive and relay their experience. We become a part of that moment and of the experiential assemblage (*sensu* Deleuze, 1994). The impact of empathetic bonding among social groups resonates with each community introduced in this book in the most varied ways. In reflecting on the importance of social support networks and empathy in memory work, I also consider the need to integrate more fully the social and the cultural in our professional outputs.

Outputs

A note about outputs. After all, to research means to publish, and publishing a piece of work in a scholarly context we attempt to 'visualise' a cultural framework in order to share knowledge we have co-created with communities. Alongside the power and workings of the imaginative collective and individual mind, we engage with the contingent issues of interpretation, representation and dissemination. For instance, how does a reflexive researcher go about making sense and even visualising entanglements of facts, data, imaginaries, changes of heart, gossip and art that make up communities' attachment to place and links to their pasts, presents and futures (see also De Nardi, 2015)?

The way to go about this is, I think, to trust publications to do something with and for the communities we work with, even if ephemeral and partial. In community research, we will never have the full picture. We can at least ensure that the community members are the main agents in our precarious experiments. Increasingly we publish and visualise not only in the interest of academic transparency and honesty but to empower communities to bring their knowledge and identity politics into a broader arena: scholarship, the digital world, international recognition and so on (Dedrick, 2018). The notions of "intersubjectivity, countertransference, and the dynamically interacting forces of projective identification, introjection and incorporation [. . .] dissolve the boundaries of self and are 'other'. They are difficult to teach effectively, and have to be learned experientially" (Barbour, 2016: 96). Heritage fieldwork can benefit from a toolkit for experiential learning and should strive for the communication of co-produced, negotiated values and knowledge before anything else.

Stories, memories and their telling are mediated not only by the cultural context in which they are produced and enacted but also by the body of the tellers and the gestures of the group (Pink, 2009). We should be mindful that research interviews and storytelling act as embodied exchanges of recollections during which information is shared but stories and impressions simultaneously occur, affect and are affected by our reciprocal presence. The strategies, tricks and dynamics through which we, as researchers, come to frame, represent and ultimately transform events and places through our intervention are as varied as the multitude of cultural groups and human societies we come to work with. And still, the multisensory experiencing of the past in the present is usually lost in official accounts and academic outputs; as experientially leaning fieldworkers and scholars, we may have to live with this fact. We can acknowledge with an amount of healthy frustration how publication 'flattens' our and our co-researchers' experience in the field. On a positive note, by engaging with communities through mixed media we can go some way towards opening up occluded layerings and linkages of memory-presents. For example, we can make this happen by inviting co-researchers to record, to draw, to map, to take photographs or simply to wander around talking to each other (see Seremetakis, 1993; Dedrick, 2018; Pink, 2009). Local actors' perceptual channels and agencies can then come to the fore, above and beyond our 'learned' thick descriptions.

The maps in this book only hold temporary meanings: their significance has changed since this book went to press. It is inevitable and yet promising. The maps are boundless. The potential of the mapping practices outlined here is as open-ended as life experience itself; the process is made of, and reflects, multiple affectual linkages, experiential assemblages and imaginaries. As I type now, of course, the book has not yet been published, which is why it is with a sense of expectation that I can only speculate on the outcome of these experiments.

The maps themselves

This book introduces completed fieldwork projects based on collaborative mapping (the three Italian case studies) and a project in-becoming (the memory mapping in collaboration with Beamish Museum in England at two sites: Ryhope and Kibblesworth). These five visualisation experiments may shed light on the open-endedness and collaborative nature of cultural heritage production. These examples embody highly diverse understandings of, and engagements *with*, material culture. I happened to become involved with the different locations and communities present in this book at different times in my life and academic career. In some cases I was an insider, mingling with my own; in others, an outsider looking in. The deep mapping methodology developed as a way of 'feeling in place' together with others during fieldwork, initially as a medium to express autobiographical insights captured during research in the field, but then developing into a gathering of agencies and affects shared with others (MacKian, 2004). Eventually, the method became a methodology with which to experience the past, present and future of place. In this spirit, this book offers practical strategies for capturing some of the essence of place as the framework for the real and the imagined, the remembered, the present moment and anxious, perhaps impatient futures.

Is it even worth attempting to visualise place, memory and the imagined? A map, although at some point realised and visible to persons other than the mapmakers, is not prescriptive in light of its transient materiality. It is vulnerable, and it is objectionable. A map with many makers is a visual storytelling rich in sensory, more-than-textual, affectual textures. I have successfully used this mapping experiment, alongside other ethnographic methods, to facilitate a group's wish to visualise the sense of place and the plurality of affects that circulate at heritage sites, at historic landscapes or in the urban fabric.

Open-endedness was key from the start, as the maps led me to consider a wealth of possible pathways and openings for this affect-facing heritage research. The experiments and the thick stories and counter-stories told *in place* shaped theoretical insights, making unpredictable snags in the frameworks I came to think with and through. The process of collaborative map-making disrupted most of what I knew about these communities. I often witnessed the instability of the cultural-mnemonic assemblage whose unexpected tilts and patterns became part of the relationship between people and place. Exchanged stories and secrets merged with the strokes of a drawing pen. I embraced the textural inquisitiveness of the visualisation method, which became so absorbing as to completely blur the line between research and experience even in unfamiliar contexts like the northeast of England and, later, Swat Valley in Pakistan. The maps usually shape my research questions, rather than solidifying the questions into the visual medium.

In practical terms, the visualisation experiments and mapping strategies I introduce here may prove useful in complementing and even enriching traditional ethnographic strategies of field observation. Bottom-up visualisations may provide more nuance to autoethnographies in the field (Wall, 2006) by positioning our agencies and those of others in the same affectual spaces. This kind of collaborative mapping may well insert itself as novel practice in the toolkit of participatory research across disciplines (Hardy, 2012; Mulhearn, 2008; Onciul et al., 2017). The multivocal and informal nature of the visualisation exercises as presented in this book reflects the spontaneity of much community-based expertise and the affectual roots of many a local historical society (Orange and Peters, 2011; Shanks, 2012). For instance, the visualisations may illuminate ambiguous emotional links between dwellers and region (or site); they may also elucidate the experiences of inhabiting spaces dense with histories and stories, memories and imaginaries that animate place. For heritage scholars and field researchers, the mapping experiments may prove to be an informal and often playful avenue to explore place meanings, real and imagined. The process of feeling, observing and mapping experience together, as a group, can then be useful to make room for hard-to-define, more-than-narrative moments that we live in everyday life and fieldwork. After all, working with any given community entails an engagement with layers of perception and the fragile interplay of private and public lives – any community is "an elusive ideal" (Keller, 2003: 9) unless we patiently and respectfully work through its many aspects past and present, accepting the worth of the memorable and the forgotten, the proudly displayed, the hidden and the hurtful.

The 'thick' (or deep) memory maps present in this volume and first articulated in a paper (De Nardi, 2014a) are not simply a colourful metaphor for memory: they also make spatial connections meaningful. To visualise something is to own up to it, to declare 'it' present even if intangible, hidden or invisible. The act of mapping out, of drawing to visualise something, brings out things that are meaningful, even if they are painful to verbalise or ephemeral and hard to quantify. Often, making the present-past visible means mapping out meaningful places and things that have disappeared before they fade from memory. In some cases, this effort entails the presencing of invisible traces; in others, the mapping process is a means of materialising loss, of making loss evident. Are visualisations a potential alternative model to individual remembrance narratives? Are they a gentle incentive to bring forth something or to let go of something else? Either way, storytelling in visual form (or storyboarding, even) may be able to build on, grow from and reframe a politics of memory to create a reflexive and more inclusive platform for future-making. I explore these implications further in Chapter 2.

Road map to the book

This is not a memory-studies treatise. This is not because I do not find the notion of memory endlessly fascinating and even compelling – I do, and memory (and its related process of *postmemory*) is ever-present in the stories unfolding throughout the book. What I want to do here is slightly different, although it hopefully adds a methodological 'something' to the debate on 'affecting' memory by facilitating the visualisation of some of memory's aspects, such as the imagined, the constructed or the fabricated memory. In working through the stories in this book, I privilege the creative side of social memory, that is, I observe what memory can do and *become* when visually expressed and collectively decanted in the context of community fieldwork and workshops. The maps leave traces of themselves as material culture items even after eyewitnesses and social actors elect to move on, pass away or lose interest. Often, one map is the start of many follow-up experiments, as is the case for several schools in Italy and the Swat Museum in Pakistan, trying out the method in novel ways that render the process their own.

The ideas of 'place' and 'trace' are closely tied to the rituals and practices of remembrance enacted by the five groups of co-researchers present in this book. As this volume also seeks to act as a guide or companion to the mapping methodology it illustrates, I provide summary bullet points at the end of Chapters 3, 4, 5 and 6. These summary points articulate the steps taken in the devising and making of the maps and visualisations, as well as go over the main ideas and arguments proposed in each chapter. In the Appendix, I provide a step by step breakdown of the mapping methodology in order for the process to become easy to follow and, if desired, implement in other fieldwork.

In Part I, based on three Italian case studies, I reflect on my decade-long work with Archaeoclub Gruppo Archeologico del Cenedese (GAC) and with two local historical societies, ISREV and the Istituto Storico Bellunese Della Resistenza e Dell'Eta' Contemporanea (ISBREC). These groups are, respectively, a Vittorio Veneto and environs-based avocational archaeological society (GAC) and two non-governmental organisations dedicated to the research and memorialisation of the anti-Fascist partisan struggle in northeast Veneto during the Second World War. Mapping became a part of the fieldwork with each community and non-professional historians as if growing out of a desire to not only investigate time, but also to interrogate the persistence in the present of spaces and places where atrocities and violence took place during the war.

In Part II, dedicated to the northeast of England, I recount the process of collaboration together with the Community Participation Team at Beamish Museum in an experiment with memory visualisation. Beamish Museum, myself and two local communities joined forces to craft memory maps of the 1950s which will be exhibited as part of Remaking Beamish – the big 1950s development due to be completed in 2021. One chapter in this section (7) includes a contribution by the Beamish Museum's community participation officer, Geraldine Straker. The reflection part of the chapter is an informal and to an extent dialogic assessment of what it was like to embark on this particular kind of collaborative fieldwork and reminiscence work. Geraldine and I articulate our collaborative endeavour side by side: we reflect in turn on how we listened to community voices to ascertain how they wanted the 1950s to be represented in the museum and beyond. The reflection section of this chapter outlines our attempt to visualise, through deep maps, the ways our co-researchers wanted their communities to be remembered ahead of the museum's 1950s town extension.

The book title's self-appointed 'visualising' mission is eclectic, so different chapters explore or foreground different contexts, themes and priorities which do not necessarily speak to one another. Similarly, the outputs and processes that make up the various cases studies may be of interest to a varied spectrum of practitioners. For instance, the deep heritage map presented in Chapter 3 may be of use to cultural heritage practitioners and archaeologists who wish to explore ways of participating in representation and ownership of cultural and historical assets, tangible as well as intangible. This case study may be of benefit to scholars and those who are interested in moving beyond top-down representations and would like to encompass community views into their outputs by 'thinking' about place and memory in a more imaginative and even playful manner.

Chapter 4 explores the urban memory of the civil war in an Italian town during the years 1943–1945. The production of emplaced, 'mobile' memory maps of the war as experienced in place by town residents across generations may be of interest to urban researchers and those interested in how history and memory work in motion: the visualisations in this chapter occurred during and after experiments in walking and embodying the past in the places where 'it' all happened – in the field, as it were – as opposed to the more usual interview context in a static location (Drozdzewski and Birdsall, 2019).

Chapter 5 pieces together through communal effort and togetherness the fragmented agency of a non-place, Bus de La Lum, a contested Second World War grave site. This chapter tackles an apparent impossibility – that of mapping experiences tethered to a non-place or a negated place of memory. The intense, fraught process of excavating and representing the unpresentable – fratricidal wartime killings – may be of use to scholars of civil war and the disappeared across Europe and beyond or to those whose research deals with what clinicians have termed 'ambiguous loss' (Boss, 2000, 2010). This site became a "memory they buried without legal burial place" (Abraham and Torok, 1994: 141). Any effort to visualise or even imagine this non-place would entail a leap of faith that may not be easily accomplished. With hindsight, the physical 'bringing together' into a room of persons who remember or 'feel' the pull of the mass grave has accomplished more than individual interviews and solitary exploration ever could.

In England, Chapters 6 and 7 offer a foray into the social memory practices of residents of Ryhope and Kibblesworth, two former coal mining English villages at the edges of Sunderland and Gateshead respectively; the stories therein complicate the selective process of museum 'message-setting' and narrative strategies, opening up the floor to the small-scale and proudly personal memories of groups. The English case studies may prove of interest to cultural historians of post-industrial communities in understanding bottom-up politics of representation in a colourful, lively and inclusive manner.

The concluding chapter then sums up the main ideas presented in the book, trying to pull together strands and threads of lived experience, both visible and invisible. After a review of the individual case studies I attempt a calibration of the mapping process, measuring the successes and failures of the experiments, and argue for the usefulness of visualising the affectual density and unpredictability of emplaced heritage and memory.

Finally, the Appendix outlines the stages through which the visualisation and mapping projects take place, drawing on the case studies presented in the preceding chapters and on best-practice observations learned through trial and error in the field. The Appendix is intended not as a guide or a manual, but more broadly as a loose breakdown of steps, activities, decisions and practices that make up the wider 'ensemble' and process of the 'final maps' as presented in this volume.

Memory in the present context

Structures of feeling, networked affects, power struggles and collective memory all compose the daily entanglements between past, present and future as they are lived and shared by communities. In this book, the decision to 'stick with' certain versions of the past validates the identity of a subgroup (the avocational archaeologists; the descendants of the coal miners; the offspring of the civil war combatants): this identity is deeply affectual rather than territorial, but the emplaced element in this tethering and attachment is given by affectual links to a 'belonging together'. This emphasis on belonging and sharing ties in with Margaret Wetherell's (2012: 143) critique of Brennan's conception of affect as something which is transmitted unilaterally – from A to B, without external deviations or interference. The problem with this model is the intrinsic individualism that other scholars of affect refute (2012: 143). Moreover, Wetherell reminds us that affectual transmission is not about inert recipients awaiting instruction – it is an infectious energy, unpredictable, that can catch us unprepared as it circulates. Drawing on Halbwachs, Connerton (1989: 20 ff.) dismantles the analytical separation of individual and social memory as meaningless. To consider the formation of social memory, it follows that one must consider how those memories are constructed and conveyed through such commemorative ceremonies and imaginative acts.

The heritage field has embraced the challenge of building stories on memory – albeit usually of a more official, visible and tangible nature (Newson and Young, 2017: 6). Much work has been done on the multiple intersections of memory and heritage (Waterton, 2005, Dicks, 2010; Sather-Wagstaff, 2011; Orange, 2010), memory and materiality (Seremetakis, 1994; Saunders, 2000; Osborne, 2001; Renshaw, 2011; Sather-Wagstaff, 2011) and memory and history (de Certeau, 1984; Ricoeur, 2004; Cubitt, 2007). Waterton and Watson, following Smith (2006: 3), qualify heritage as a 'process' "linked to memory, identity, politics, place, dissonance and performance" (Waterton & Watson, 2010: 4).

Overall I feel that, in the current post-factual economy of the 21st century, the imagination and the make-believe play a central role in how people, structures and institutions come to think of themselves, each other and other actors in the social sphere, augmenting if not creating new memories in the process. If someday cultural ethnologists come to study the rise of Donald Trump to the presidency of the United States, they may turn their attention to the popular socially constructed fictions that surrounded his presidential campaign. They may investigate the pervasive fascination with certain tropes of home, security and protection embodied in political myths like the Mexico border wall. The future scholar's 'anecdote' is just a trope, a story among many, but it gives an idea of the unpredictable and unlikely things which come true because people will them into being or believe in them strongly enough.

Indeed, memory has as much to do with presents and futures as with the reworking or reimagining of past events and past identities (Berliner, 2005). Less canonical strands of memory may include "the incidental, the insidious the retrospective and the accidental" (McCarthy, 2017: 61). So, for instance, the Leave.EU movement in the UK sought to shape that country's national and international future outside of the Single Market and the European Union based on imaginings of future independence, border control and a return to British values of the past. At the centre of an orchestrated projection of the ideal Britain into the future was a keen sense of a past that likely never occurred but that satisfied the desires and longings of a certain nostalgic right-wing fringe within the political and social spectrum of the UK.

Across the world, the British fleet's arrival on the Australian continent in January 1788 is a key event in Australian history, one that is a haunting milestone in collective memory and constitutes a pervasive trope of colonial violence (Connor, 2005). The 1788 settlement has acquired elements of a 'foundation myth' for the descendants of British colonists and convicts, while the fateful date marked the start of an age of violence and discrimination in the Indigenous communities and the traditional countries that made up Australia prior to contact with the West. Often framed in racial and postcolonial discourses, the experience of countless Indigenous Australians has given rise to the Change the Date movement. And yet, the imaginings of these events could not be more different, as the memory and the postmemory of these landings are imbued with affectual understandings which depend on the positionality of those who remember (imagine) the eventful date of 26 January 1788. How this postmemory is framed and felt depends entirely on whose stories we foreground.

Viewing heritage and 'affect' through the optics of co-production

Mapping can be understood as a kind of more-than-representational visualisation of memory and historical experience. In a way, heritage or memory maps and visualisations have the potential to bridge the chasm between the factual and the imaginary through the vibrancy of affect. The attractiveness of the affectual paradigm – with its promise of inclusivity, imaginative world-building and embodied inquiry – has increasingly inspired practitioners to experiment with notions and practices outside and beyond the discursive (Bondi et al., 2005). The workings of affect are a tool "to elicit intense embodied, physiological responses with very powerful effects" (Sather-Wagstaff, 2016: 12). Initially, non-representational theories surged in response to increasing interest in practice and performance in cultural geography (Thrift, 1997; Valentine, 1999; Philo, 2000). For Nigel Thrift, the non-representational project sought to define "practices, mundane everyday practices that shape the conduct of human beings towards others and themselves in particular sites" (Thrift, 1997: 142); in Gill Valentine's and Judith Butler's (1997) work queering the politics of embodiment, experience of the social world is steered away from the heteronormative paradigm towards freer entanglements of matter and materiality afforded by the non-representational episteme. In accordance with the multiple and non-prescriptive declinations of affect, the implications for the spatial have been significant: the non-representational canon reinstated the legitimacy and freedom of the sensing body and the boundless realm of the imagination to the geographical agenda (Wylie, 2005).

Among the notions reframing geographical ontologies in terms that exceed the representational, affect theory has certainly found the most purchase. This drive engulfed the social sciences at the same time as the humanities, often at their intersection. Affect acts as a set of embodied practices that make visible conduct an outer lining (Thrift, 2004: 60), as intensely social as it is political – in the same way that heritage practices and the imaginaries that affects inhabit and reanimate are political. It is possible to note the importance of affect for "the maintenance and production of memory as well and cultural and social understanding" (Gregory and Witcomb, 2007: 263) as time and place coalesce in affectual assemblages. Affect studies do not just probe into the experience of the liminal or extraordinary, but also foreground the seemingly humble quotidian practices that make sense of the world (Bondi, 2005): for Kathleen Stewart, the realm of the everyday is where affectual encounters are most pervasive, encompassing a myriad experiences that form memories and shape our sense of self, place and identity (Stewart, 2007).

More to the point, the momentum in studies on affect and materiality has served to bridge the gap between the abstract idea of memory and elements of heritage inquiry and practice (see Trentmann, 2009). A thrust towards inclusion of the small scale in our research has raised the bar. The acknowledgement that any utterance may be posited under our academic lens as an event imbued with the workings of social and learned practices allows us to embrace the occurrence of ethnographic events even when they manifest as spontaneous, unrehearsed affects (Tolia-Kelly, 2004b). We no longer discount the significance of a Proustian moment of recall triggered by the senses before the conscious mind has the chance to catch up, cherishing instead the vibrancy of the encounter with our co-researchers.

At the same time, tetherings to the observable material world continue to sharpen our analytical lens. We may witness moments of illumination and sudden recall but are mindful that such serendipitous occurrences are often linked to the cultural and spatial contexts within which the subjects we work with operate.

Projects of affectual reanimation of heritage and imaginative participation have emerged under the rubric of the so-called material 'turn' in the social sciences (Gregory and Witcomb, 2007). Scholarship has increasingly turned to the workings of affect in non-representational theories and beyond, leading to a lively critique of notions of identity, place, emotion and embodiment in the social and spatial sciences (Witcomb, 2010; Sumartojo and Pink, 2018). A surge in interest in storytelling (McCormack, 2003, 2010; Anderson and Harrison, 2010a, 2010b), a repositioning of embodiment (Wylie, 2005), a focus on the everyday (Stewart, 2007) and the unmasking of affects as a socialised politics of emotion (Ahmed, 2010) characterise much of this literature. Nowadays, the poetics and impact of affect are seen as increasingly relevant to (if not embedded in) the practices and theories of heritage and museum studies (see Witcomb, 2010; Hall, 2005; Turner, 2016; Sumartojo and Graves, 2018 among others). Several recent studies problematise the very nature of heritage itself in assessing its effects in the present through the politics of the past (Tolia-Kelly and Crang, 2010; Waterton and Dittmer, 2014). The circulation of affect is a powerful paradigm to interpret the politics of the present too: motivations, social and cultural constructs and performances afford different heritage trajectories across a variety of contexts. Affects transpire from actions and materialise intentions, shared by more than one agency (Hall, 2005). Heritage is understood as operating within a wider affectual assemblage insofar as its practices, dynamic and engulfing, operate not as encoded discourse but as a form of social groups' historical imagination (Muzaini, 2015; Crouch, 2015) – often open-ended and unstable, but always valid.

Traces of the past are all around us, in this world we come to share with others through our engagement. "Our connections to the past", argue Waterton and Watson, "are [. . .] tangible, and [. . .] have a materiality upon which they depend that makes them objects of heritage" (2010: 2). Things resonate; place meanings ebb and flow in time. So, for instance, the rejuvenation of neglected places and spaces by a deindustrialised community is felt as a presence much in the same way that the hollowing out of that very community's sense of purpose carved out an absence of activities and things (Walkerdine, 2009; Orange, 2010). Heritage can thus help animate and foreground presences and absences, championing these visible and invisible forces for the benefit of communities (Orange, 2017), so much so that heritage perception, like memory, can bridge the material, the social and the affectual. When undertaken in respectful ways, heritage practices may even act as a conduit for values variously tethered to place, memory or imaginary (Smith and Campbell, 2011).

Indeed, affectual exchanges may exceed or even upend the expectations or preconceptions we may hold about the way heritage works (Smith, 2011). Heritage as a dialogic, open-ended affectual assemblage engenders practices so diverse that even painful experience leads to a tethering or filtering of reminiscence or recall in a specific spatial context (Tyner et al., 2012), so much so that "place attachment is a key concept in heritage work, but there is strangeness in the idea of being attached to things that hurt us" (Byrne, 2013: 605). The spectrum of affectual engagement in heritage is thus vast and its experience sometimes wholly unexpected – or unwelcomed. Some locations may activate or channel hurtful encounters with an uneasy past (Witcomb, 2015a, 2015b), where 'reaching out' to earlier periods via affectual processes can be difficult to negotiate in what Joy Sather-Wagstaff calls "dialogic memoryscapes" (2011). In the presence of hurtful emplaced memory, powerfully evoked by the presence of the sensing body at the site of a tragedy, the occlusion of affectual communication between past and present complicates the heritage event in its unfolding. However, Sather-Wagstaff has also postulated that even "dark heritage" encounters enable a multiplicity of affectual openings (Sather-Wagstaff, 2016: 13). The affectual spectrum may then be expanded to include heritage encounters that stimulate forgetfulness, denial of memory and dissonance. For Barbara Voss, "artifacts can evoke powerful responses that draw emotional connections between the present and the past" (2018: 540). Undeniably, responses to evidence of suffering are not always a given, and the context of commemoration and the nature of the heritage encounter can either enable or disable empathy and imaginative 'closeness' to others (Witcomb, 2015c).

Besides informing research across the board, the politics of affect is central to the ethos of collaborative mapping. Affectual understandings of heritage (open-ended, inclusive, multiple, irrational, conflicting) expand our ability to work with others in building and shaping sense of place in ways that make sense to the communities we work with and for. First, we may reimagine the affectual process of heritage as something dreamed up, evoked and shared, made and circulated by communities (MacKian, 2004). Heritage does not and cannot operate in isolation, and museums have quickly grasped the implications of this synergy. Different modes of learning about heritage emerge in relation to specific contexts: thus, "the perception of intangible traces of past life incites a range of emotional and cognitive experiences" (Gregory and Witcomb, 2007: 265). The shared perceptual experience of being at a site of historical significance acts as a social experience insomuch as it is intimately moving and pervasive at a local level.

As noted, the linkages between heritage processes and the realm of the affectual are manifold, encompassing the spectrum of encounter and experience from seeing an object, entering a place with a past or talking to others about their experiences and memories (Sumartojo and Graves, 2018). Wetherell's definition of affect as "embodied meaning-making" (2012: 3) can be translated in heritage terms as co-production through the senses and the imagination. Affect may also entail "relations practised between individuals rather than experiences borne by sole individuals" (Richard and Rudnyckyj, 2009: 61).

The various channels and encounters through which heritage emerges as a practice are as dynamic and eclectic as social groups and end users of heritage, so much so that conduits of heritage experience like museums have been repositioned as affectual foci of community learning (Witcomb, 2015a, 2015b). Thus, an engagement with things and places of the past framed through shared encounters may disclose an array of political and social practices that enable (or hinder) learning and participation. And so, "museums developed various ways of privileging the voice of experience rather than abstract categories of thought such as the nation" (Witcomb, 2015a: 326). Affects channel the 'voice of experience' in a pervasive but often imperceptible manner: in the everyday, intangible moments may be captured and enabled by shared experience (Ahmed, 2004). When shared stories are not at the forefront of a heritage site's 'essence', the affectual tethering may be harder to detect or nurture, which is why most museums and archaeological sites often lack the immediacy of the spontaneous affectual encounter (Tolia-Kelly, Waterton and Watson, 2016).

Further, we might consider the heritage experience as a process variously shaped and shared by publics, practitioners, re-enactors, local residents and other stakeholders in 'making' place (Dicks, 2010). Affect thus works through heritage co-production in shaping past understandings; 'affect-as-co-production' operates as a connective tissue between remembered and imagined things that matter in people's lives – things they may share, or hide, but nonetheless deem important in defining their place in the world (see also Mulhearn, 2008). The workings of affect in heritage co-production and memory work, emerging as performative storytelling practices, inform my (the fieldworker's) understandings of place. One thing is certain: heritage practitioners will never be short of stories to work into narratives through the process of co-production (Kavanagh, 2000; Cashman, 2008). When it comes to heritage experience, every community is unique, and so will be the modes of co-production at stake. In most cases, heritage may be presenced as an assemblage of shared experiences, affectual connections and memories that are variously tethered to collective place attachment (Onciul et al., 2017; Tolia-Kelly, 2010). The idea of an interrelated perceptual world made up of multiple agencies is fundamental to this book in more than one way: first, because it removes the barriers between culture and nature; second, because it transcends the dichotomy between people and material culture in its analysis of subjectivity and agency; last, because it proves the impossibility of dividing 'fact' from imagination from the workings of affect and memory in the social sphere.

Affect and the agency of the body

The affectual and non-representational canon draw on tenets of embodiment and perception well established in Continental philosophy but expand its reach by plugging in contemporary concerns and concepts to its core framework. The idea of affect itself arises chiefly out of the phenomenological tradition, attentive to the embodied dispositions of environmental actors in the production of place and culture (Merleau-Ponty, 2008). As a long tradition of phenomenologists have postulated, we exist via the medium of our sentient bodies. Langer summarised this point eloquently when he wrote that our body is "inseparable form the world of [its] perception . . . I perceive always in reference to my body" (1989: 41). The main flaw with phenomenological preoccupations was a focus on the body universal, linked to problematic issues of Western identities and ahistorical and apolitical embodiments. Situating instead the idea of affect as embodied and an intrinsically sexed and raced phenomenon socialised and experienced in the everyday serves to open up its potentialities (Ram and Houston, 2015). The associations between affect and the body 'meet' across disciplines to explore ways that embodied, gendered, political, raced social beings feel and understand 'things'.

The concepts of embodiment and emotion thus afford a fruitful framework for knowing the world in cultural and historical research. A vast literature links up the notions of embodiment and the sensing political body with the phenomena of affect and emotion. Among key works on these intertwined methodological 'manifestos' foregrounding the body and emotion in making sense of the world are the works of Stoller (1997), Bender (2006), Stewart (2007) and Wetherell (2012). Paul Stoller's phenomenological ethnography seems particularly relevant to my thinking on the imaginary and 'imaginative' experiences of the past, as it foregrounds the intangible elements that pervade the body social through the experience of storytelling. For Stoller, "even the most insightful writers consider the body as a text that can be read and analysed. This analytical tack strips the body of its smells, tastes, textures and pains – its sensuousness" (1997: xvi). He makes a compelling case for the body to be subject not object of ethnographic encounter and analysis.

For Barbara Bender, in order for humans to make sense of the world and their surroundings they must perforce engage multi-sensorially with it. Being in place is not a matter of visual perspective, but a series or web of "intimate encounters" (Bender, 2002). Dwelling or making sense of place is not about seeing, but about experiencing a somewhere through the senses (Bender, 1993b, 2002). Much like Stoller and Bender, Wetherell vindicates the centrality of the flesh and embodied agency in the affectual turn. For her, a focus on affects should be shifting attention away from discourse, attempting to recover "the energetic, the physical, and the sensual" (Wetherell, 2012: 9). And yet, a Eurocentric bias in the turn to affect may exist: in vertical societies oriented along different embodiments, not all actors may be able to express and partake in public acts of affectual exchange. Most traditional affectual autoethnography will be framed through a dominant set of historical and racial parameters that may not be inclusive of other subjects, and much cultural affectual literature may still negate the silencing effect of institutionalised violence towards subjugated and marginalised subjects. For Tolia-Kelly, it is essential to also attend to a "body-centred approach to identity. [. . .] The body in a situated experience of landscape is what is pursued through ethnography" (2010: 5) in a wider and more inclusive sense.

An attention to the small scale may go some way towards rendering affectual notions of the sensing body more inclusive. Stewart (2007) instead focuses on the enchantment of 'everyday affects', the multiple textures of our lives that we seldom stop to notice, entrenched as they are in our feeling of being alive. Our bodies have become so attuned to these non-human elements in our lived environment that although these other-than-human elements may blend into a harmonious background, they can also startle, harm and disorient (Maantay, 2013). The range of affectual expression can span "big moments" as well as "more banal and everyday experiences, some of which may well be fleeting, equivocal and muddled" (Wetherell, 2012: 43). Affects, after all, can be "funny, perturbing, or traumatic" (Stewart, 2007: 2). In the presence of trauma, the dialogic memoryscape is compromised: what happens when the articulation of stories and events *in place* present a defensive 'closing' of perceptual doors, instead of creative, outgoing affectual possibilities? Chapter 5, in particular, presents just how disharmonious, melancholy and indeed equivocal affectual registers can be.

The roots and positioning of affect theory within the logics of phenomenology may help unpack this recent metaconcept as part of a framework to make sense of the multi-agential world – and of the past as experienced in the present.

> On an agential realist account, it is once again possible to acknowledge nature, the body, and materiality in the fullness of their becoming without resorting to the optics of transparency or opacity, the geometries of absolute exteriority or interiority, and the theorization of the human as either pure cause or pure effect while at the same time remaining resolutely account-able for the role "we" play in the intertwined practices of knowing and becoming.
>
> (Barad, 2008: 139)

Any aspect of human experience is grounded in the sensory assemblage of the world, a world always in becoming in which multiple realities converge. Karen Barad's theory of agential realism serves to gather together the material and immaterial, the human and more-than-human, as interlinked agents of experience and sources of memory, interaction and the social. Framed in material feminist concerns, 'affect' becomes one of many agencies in a complex assemblage of energies and entanglements that make sense of the world. And yet, admittedly, humans are the ones making (and using) the kind of maps and visualisations described in this book. Human interactions have shaped places and led to their framing in memory and the imagination – for the most part. Maps are discursive and narrative outlets of representation: signs on paper or on a computer screen. That may be, but the agencies, the temporalities, the affects that make up place and are then dreamed up and spatialised in the visualisations are the result of a concerted 'effort' by various agents in place (see here DeSilvey, 2006). We are not banishing narrative and representation to the doom of forgetfulness; we are not tricking the mind through a focus on the senses to stop seeing maps for what they are: signs for something else. Here, semantics are not of essence in a process whose focus is the journey rather than the destination. And after all, visualisations can never be fully Cartesian if they project the invisible and impalpable onto the visible fabric of place. On the whole, we should not attempt to understand mem-ory and human experience of the past without weighing in the burden (or gift) of the embodied imagination as a force shaping everything: past, present and future across multiple agencies.

Temporality and the 'porosity' of stories

A discussion of emplaced heritage, memory performance and identity tetherings necessarily leads to a reflection on temporality. The temporality of landscape and place is yet another actor in a mesh of affects existing in a location and permeating its atmospheres (Sumartojo and Pink, 2018). Time-place evocations also have a peculiar politics through the workings of remembrance (Schama, 2005; Lowenthal, 1985). Encounters with the past are often encased in the sensory experience of vicarious loss and mourning; thus Witcomb (2010) details how the exhibition of a Treblinka Nazi death camp replica built by a survivor leads museum visitors to vicariously experience loss and bereavement through the construction and display of this object. Such reframings of sense of place and temporality through the prism of memory give memory a new lease on life, steeped in the affectual and post-human (DeSilvey, 2006; Maantay, 2013).

For Yi-Fu Tuan (2004: 53) "place . . . slows down time, stops it and can even, by offering the possibility of return – reverse it". But can time ever really stop? Is it not more likely that we can grasp or capture a moment in time in order to make sense of it – and blow it up, enlarge it and analyse it? The map-like visualisations I discuss in the following chapters can accommodate incongruous views of the past in their fold: that is the point. They portray experience as it intersects time and place. This cross-pollination marks experiential assemblages, its open-endedness of time allowing for an expanding affectual spectrum encompassing past, present and future energies and agencies (see also Andrews et al., 2006). Perceptual shifts redefine local geographies across the experiential realm in ever-changing ways (Blum and Secor, 2014; Light and Young, 2016). As Susan Hekman postulates, "as the elements of the assemblage change, so does the realm of practice and experience" (2008: 100). This book notes how relevant this idea is to the complex materiality of the Italian civil war after the shift in allegiance from Nazi Germany to the Anglo-American allies. Shifts in the experiential assemblage and environment (Robb, 1994) have also adversely affected communities and affectual economies in the aftermath of coal mine closures in Britain (Strangleman, 2001). Entangled in nature and culture, time is just one of many agents in these evolving assemblages of loss and renewal.

The imagined past and the reality of historical events may be perceived as one and the same, a poetics shaped by and in turn shaping the lives of social groups (Black, 2003). It is often the case that traumatic memories are the ones that get remembered the most 'unevenly' across the board. Memory fails, and the imagination takes over out of necessity or choice. Chapter 5 shows how this fragmentation works in practice. Why should time be mapped as manifested through its entanglements with memory or the imaginary? If temporality is but an element in the affectual assemblage of memory, deep mapping blends and incorporates elements of landscape that intersect time as well as space. Perceptions of the deep past also complicate and enrich our notions of place, memory and perception especially if we move past the Western perspective on temporality as fixed. Thus, as Ann McGrath has argued (2015, email communication),

'historical' scholars (as opposed to archaeologists, prehistorians) may contribute to "humanizing the deep past" by engaging with the stories of individuals, such as Lady Mungo, who lived at least 40,000 years ago.

As revealed in the film *Message from Mungo*, it was easier for White Western participants to narrate Lady Mungo's role in the present rather than in the past, whereas the film's Aboriginal participants challenged Western notions of temporality in portraying the vast time gap between Lady Mungo's existence and their own as relatively meaningless; to them she was "like an old Aunty who died yesterday" whose living presence continued to intervene in the present. Until she was returned to Country, Lady Mungo was a displaced individual in the Aboriginal perception. This serves to demonstrate that long time becomes intimate time if the prism of its observation is changed: in that case, place-tethered affects bring chronologically distant time that little bit closer.

Time and memory work in synergy, informing one another and entangled in the same experiential assemblage. Scholars from as varied disciplinary backgrounds as psychoanalysis and oral history (Conway, 1996) have argued that "the memory process depends on that of perception" (Thompson, 1998: 129). For Ricoeur, "our very inquiry into the function of narrative [*shapes*] our temporal experience which will compel us to go beyond the obvious but provisory opposition between the *direct* truth-claim of history concerning past events, and the *indirect* truth-claim of fictional narratives which are implied in the various forms of the mimetic function" (1991: 105, emphasis in original). Moreover, cultural memory is "never left behind, even when exploring unfamiliar places. [. . .] All landscapes are a product of myriad decisions, each mediated by place imagery" (Blake, 529).

Decisions are stories, or rather, choices about how a story ends. Their open-endedness defines stories even when a version of the story is at some point chosen as the 'final' draft. In Italo Calvino's self-penned introduction to classic Italian Resistance novel *The Path of the Spider Nests* (1964), the author struggles to describe why he came to write what would become an iconic European literature classic and a Resistance literature milestone. In his brilliant, artfully faltering introduction, the author kept changing his mind and starting over. At one point, Calvino (1964: xi) pinned down his unease with the manuscript's final draft in one sentence that, alone, seems apt to capture the *very* essence of memories and legacies of the wartime experience:

"There were so many stories".

But whose stories? How are these stories captured and shared? Take wartime Europe as an example. The social and cultural relevance of the Resistance is ever more prominent nearly 70 years after the end of the Second World War: it is an iconic moment in Italian history and identity (Neri Serneri, 1995). And yet, communities that were affected by German occupation, Fascism and Resistance activities have still not made up their minds – not only about what happened back then, but about how they *feel* about what they *think* happened in their hometowns, neighbourhoods and streets. German reprisals in Italy and France made the situation even worse for any civilians suspected of collaborating with Resistance networks. Now, we still do not comprehend the full impact of the German occupation on civilian morale, perception and imagination, and we know little of how

German soldiers perceived and interacted with the landscapes and people after Italy's capitulation, when the once-allied soldiers found themselves suddenly thrown into a new hostile *Umwelt*. A once safe haven became a maze riddled with secret hideouts, ambushes and acts of sabotage carried out by crafty Italian partisans. And Italian civilians, who did not usually understand German, often presented a resentful united front, mute in their understandable terror, and thus contributed to the sense of alienation felt by German soldiers (Focardi, 2013; De Nardi, 2016). In wartime Europe, invisibility and silence 'are' the story. Around each corner, behind each window, there could be an enemy, watching and waiting. Seldom have these been framed in a perceptual or experiential line of inquiry – let alone visualised. The histories of the war in Europe thus become the experience of conflict itself (van Boeschoten, 2005; Cappelletto, 2005).

In encountering difficult and often intangible heritage, we often need to turn to the embodied, mnemonic and non-verbal for clues. As Rachael Kiddey postulates, conceptualising "homelessness as heritage is counter to prevailing notions of heritage as that which is 'desirable' or 'worthy' of preservation" (2014: 177). Framed in the present through the pervasive process of materiality, even invisible or marginal social and economic processes can be better understood. In the positioning of heritage, voices and silences are alternatively and dynamically articulated along multiple timelines: memories, emotions, postmemories passed down from one generation to the next are part of a brecciated temporality of experience (see Chapters 4 and 5; see De Nardi, 2019). In my research, temporality, intended as the 'wartime' lived and experienced now and during the war itself, is meaningfully shared across generations of people and families. Sometimes, in memory, time stands still, as if people were afraid to shatter a fragile object by resurrecting the dead. Often, however, the timescale of memory of the war transcends generations through communication and the sharing of stories. For Marianne Hirsch, postmemory is a kind of "living connection" between generations (2008: 104). Thus, in post-industrial Britain the old and the young rehearse shared memories of the 'pits' and forge new ones, together (Walkerdine, 2009). Moreover, this process of active 'making' resonates with the primacy of stories over history: everyone has one to tell and share. Places and things are part of the same entanglements, caught up in a world of stories which can be pleasurable or painful to tell (see also McDavid, 2002; Sather-Wagstaff, 2011; Tyner et al., 2012).

In the chapters and sections that follow, I begin to trace some of the insights and imaginations that colour and shape stories about place across two regions in, respectively, Italy and England. The different ways in which to approach collaborative 'heritage' mapping in different contexts can provide some useful terms of comparison for other researchers or educators doing work with "mnemonic communities" (Olick, 2008).

1 Defining terms

What is place?

This chapter articulates some of the core themes in this book, starting with the metaconcept of place. "In memory, time becomes 'place'", contends Alessandro Portelli (1997: 32). Place can be built up, revisited or discovered anew; the very idea of place is imbricated in a constant process of becoming (Belcher et al., 2008: 501). The many constituents of place could be disassembled and analysed, but, if one were to attempt to piece them together again, they would not achieve the same configuration as before – least of all would they find themselves in the same place.

Place is certainly one way that remembrance is attached to the visible, tangible world. But this is not always the case. Michel de Certeau has asked, "Of all the things everyone does, how much gets written down?" (1984: 42). Emplacement, or the situating of memory in a meaningful space, does not necessary lead to an objective or streamlined way of remembering. My main concern in this book is the ephemeral and unstable nature of stories that generate, shape and transform place. From place, to the body. For Curti, when we explore "bodies and how they affect and are affected by one another, the material conflicts bound to politics of memory and place may also be understood as political strategies to relationally re-shape and re-constitute (Other(s')) emotional geographies" (2008: 116). The performative element of 'heritage' negotiations and experience is notoriously difficult to translate and represent in traditional scholarship and (paper-based) academic outputs (Candida-Smith, 2003), yet it lends a whole new dimension to place and memory of place, *in* place.

I suggest that perhaps communities, in their imaginary bond or through shared ingroup logic, construct and experience place as so many real and imagined facets to make sense of the world and the stories told therein. They do not simply transmit these knowledges to each other as much as construct them, enact them and share them together. Community as a way of learning. Sometimes these constructions affect the tangible, the material, and sometimes they do not. The broadness and slipperiness of definitions of place single it out as one of scholarship's foundations so much so that "if two different authors use the words 'red', 'hard', or 'disappointed', no one doubts that they mean approximately the same thing. [. . .] But in the case of words such as 'place' or 'space', whose relationship with psychological experience is less direct, there exists a far-reaching uncertainty of interpretation" (Malpas, 1999: 19).

Place is hard to define as it is forever caught in a process of becoming (Cresswell, 2004); it is embroiled in an unfolding dynamic that the material feminists designate as 'agential realism' (see Barad, 2007). Place is articulated through multiple intersecting processes enacted by human and more-than-human agencies and affects; these entangled agencies are what make the spatial experience unique, ineffable and complex (Tuana, 2008; DeSilvey, 2006, 2012). Affects may attach themselves to objects as well as places and human and non-human animals (Ahmed, 2010). Beyond socialisation, place is enmeshed in a network of affects, energies and agencies that exceeds the verbal and the discursive (Deleuze and Guattari, 2004; Anderson, 2009). The situated nature of affect (Ahmed, 2004; Richard and Rudnyckyj, 2009) is connected to the shifting significance of place, to which affects are typically tethered (Watson and Waterton, 2015).

First to highlight the impossibility of place remaining unaltered for more than a few instants, rendering it impossible for someone to step into the same stream twice, was Herodotus of Samos: the Greek philosopher unkindly dubbed "The Obscure" was anything but that, for he astutely pinned down the changeable and subjective nature of place in one brief yet effective metaphor. Later in the classical world, the Greek and subsequently the Roman imagination anthropomorphised place by assigning it personhood and identity as a Genius, a spirit dwelling and dictating site-specific terms and modes of interaction between a site, humans and their lives (de Certeau, 1984; Vernant, 2001).

The ways in which space and place are conceptualised rely on several factors and agencies. Places are what they are because "people and things occupy them, give them shared meanings and situate them in collective memory" (Beauregard, 2013: 16). Place is, above all, intensely embroiled in the politics of identity (Seremetakis, 1994), and community is always (or almost always) tethered to a real or imaginary place. "With few exceptions, community always denotes a there" (Keller, 2003: 7).

Whenever we think about the power or knowledge gained by 'interpreting' a place, it is helpful to identify practices and behaviours deemed appropriate or unsuitable for those places (Tolia-Kelly, 2004a; Drozdzewski, 2015). Allowances and affordances (de Certeau, 1984) and acts of resistance to established rules offer insight into the perception of place and to the make-up of its socialised nature. Performativity and the body politics work in synergy as emplaced practices: one is the channel of the other, and they act as emergent agencies understood through the lens of relational attachments. Judith Butler (1997) has advocated the primacy of the emplaced body and of an interpersonal stance in the performance of identity and social and cultural norms under the rubric of 'performativity'. Performativity in place does not cohere into "a singular 'act', being instead the reiteration of a norm or set of norms" (1997: xxi). The ideas of resistance to the norms and conventions of a social norm find purchase in this context: the enacted memory-stories in the following chapters resonate with the concepts of resistance and performance, which help frame the ethnographic experience of 'working with' community memory.

And yet material feminists (Hekman, 2008) have questioned the peculiar bodiless-
ness and lack of 'matter' in Butler's articulation of her pervasive theory of performativity (Butler, 1993). The idea of a cultural or political separation from
nature in the affectual make-up and experience of place has also been extensively dismantled in scholarship (Barad, 2007; Tuana, 2008; Ingold, 2000, 2011;
Ahmed, 2004; DeSilvey, 2012; Bradley, 2000; Witmore, 2007). Affect permeates
the 'cultural' – whatever that means – and that which is not in the traditional
sense. As in Tim Ingold's still deeply persuasive essay on the practice of dwelling
(2000), the place-experience process does not imply any effective inertness on the
part of nature. Above and beyond environmental determinism, the natural milieu
effectively teaches and shapes human agency, and 'nature' communicates something of herself to human animals in a mutual relationship. An action, a memory,
takes place somewhere or is tethered to a specific location. Thus, place is as much
of a subject and character in any story, geographically situated or otherwise.

In contrast to the porous and fluid process of mutual learning expressed in the
dwelling perspective, place-making is a process of appropriation: it means making one's mark on the landscape. Place-making is all about imposing an interpretation or meaning onto place as opposed to revealing and exploring existing
themes, as opposed – in essence – to establishing a dialogue with nature (Ingold,
2000: 42, 55 ff.). In place-making, new meanings are negotiated. Place experience is always more than visual and definitely never abstract; place engulfs us
through the senses (Cresswell, 2004). The encounter of person and 'environment'
ultimately takes place through sensory engagement (Ingold, 2000: 56). The maps
and strategies the reader finds in this book seem to suggest that sensory and political interpretations of place are possible at the same time, through a process of
learning and making together.

Place is capable of haunting, lingering and resonating long after we have moved
away from it or erased it (de Certeau, 1984; Tolia-Kelly, 2004b). Any process "har-
nessing these placed-practices has the effect of developing social understandings
into socio-spatial understandings of knowledge" (Anderson, 2004: 257) – social
knowledge, always situated somewhere. The maps I use in this book might just
represent one such process, and these maps are certainly 'placed practices'. Placed
or emplaced practices do not just underline and nurture presences, however: they
can also mourn absences and expose wounds (DeLyser, 1999; Navaro-Yashin,
2009). Communities that once defined themselves spatially do not necessarily
need to be located in a place to nurture a strong sense of belonging. Imagination
and fantasy notions of place, long-distance longings and nostalgia for a homeland,
a town, a street, can all fulfil the need for closeness to a thing, a place, an 'idea'
of a place which may or may not correspond to an objective truth (see also Miller
and Parrot, 2009). But the hauntings, the experiences, they remain, they linger,
they are potent and exhilarating and they are contagious. The sites and memories
in this book demanded to be confronted, explored, and they asked us to get to
know them for better or for worse.

In Chapter 3, the communal imaginary of a township draws inventive, more-
than-spatial maps of a longue durée landscape through the senses and citizen

science. The resulting map is a palimpsest of ideas, impressions, memories and expectations. In Chapter 4, a town in northern Italy is held captive by the invisible traces of fratricidal violence during the Second World War. In an intangible yet haunting mesh of places and non-places, memory and postmemory intersect to conjure up a network of affects.

In Chapter 5, the dead once contained in a sinkhole deep in a forest inhabit a double identity as invisible yet haunting ghosts of Italy's fratricidal Fascist past. To visualise a place like this is an experiment in ghost photography: the attempt to capture an ephemeral, melancholy moment in time that defines a local culture and shapes a community's sense of place in relation to the outside world. In Chapter 6, an open-air social history museum sets the (staged) scene for a collective postmemory and longing for the golden age of coal mining in the deindustrialised northeast of England. The affectual engagement of many storytellers crafts new stories, striving against forgetfulness. Chapter 7 takes a dialogic form in tracing a double journey of experiments with mapping, narrated and led by an academic and a museum Engagement and Participation officer. The double vision/double journey we create in that chapter contains many overlapping layers of meaning, affects and projects.

Meeting the imagined

The lure of the imagined has long fascinated thinkers and artists alike. The notion of 'imagined communities', also critiqued by Harris (2014), was initially espoused as part of Benedict Anderson's (2006) critique of nationalism. Anderson conceptualised imagined communities as social networks reliant on the *idea* of an identity constructed and enacted by individuals – but those individuals need not ever meet in person. Their imagination provides the stuff to fill the social blanks, leading a certain 'imagined community' to feel united by a common 'way of belonging', or bound by shared ingroup logics. Geography and history, or rather the concepts of place and time, "play out in the psychic words of individuals" (Lieberman, 2015: 2); they are encased in the notions of selfhood and community, and they shape experience in all its many facets. In our imagination, we can be at one place and in many places at once. Exploring Giorgio Agamben's idea of the 'exception', Belcher et al. define it as "situated on the edge of materiality, the state of exception has the potential to materialize or not to materialize actual spaces of exception" (2008: 501). The idea of exception can contribute a nuanced layer of interpretation to the idea of the potentially disruptive, almost-but-not-quite present notion of the collective imagination. Barbara Misztal has written extensively on the configurations and grammars of social remembrance, but imagination does not appear as central to her conceptions. She does explore the idea of invented traditions, but the two 'ideas', while they intersect and mutually inform each other, are not that closely aligned or mutually unpacked.

In cultural geography, imaginaries of place loom large as subjects with an agency of their own. Bachelard's poetic phenomenology leads to framing the imagination as a "perpetual interaction between the human subject which imagines and the image itself. Imagination is thus recognized to be conscious of something other than itself which motivates, induces and transforms it" (Kearney, 1998: 97). In the introduction to their edited volume, Jane Kenway and Johannah Fahey (2009) argue that allowing the imagination to become an important and active part of the research process and project allows us to get closer to communities and to better enable the sharing of knowledge. They muse about the way that "certain people take us to such untravelled words" (2009: 1). I love this definition of the imagination: not an unreal world, but more like an unexplored territory. Being mindful of the power implications and the ever-present danger of colonising the other's imagination, framing the imagined as a potential place to discover, enjoy and explore is a good starting point to shape the 'reality' and 'verisimilitude' of memories and perceptions that seem to eschew factual events, objects and spaces. The realm of the imagination bridges the inner and outer nuances of place-making and place understanding and influences knowledge-building in myriad ways (Orange and Laviolette, 2010).

Imagination is central to this book, and, I argue, to any holistic and grassroots notion of heritage and memory, cultural or otherwise. Imagination colours pasts and presents and paints futures, shaping experience and shaped by experience and potential things-to-be. Here, I build on the idea of imagined communities and Harris's idea of community as assemblage (2014) to think about the ways communities and social groups navigate, give meaning to and remember place.

The communities I worked with told spatial stories in ways that suggest the past as an 'insistent' force in the shaping of place and in the framing of their presents and futures. This insight aligns with current thinking on identity-making linked to 'sense of place' and to multiple senses and appreciations of the past. Edward Said's conception of the 'invented' is crucial in framing place-identity configurations and national narratives (2000), as is the idea of the manufacture of heritage (*sensu* Alsayyad, 2001) as a transforming society's desire to shape place narratives and communicate selected place understandings. A drive to self-represent, making use of imaginative and creative processes to foreground one specific image of a country or regional identity, is innate to humans (Hobsbawm and Ranger, 1993; Babadzan, 2000; Bell, 2006). We often encounter this imaginative self-shaping process among and within social groups (and individuals) who have undergone dramatic change, slow decline or dispossession or who have been the victims of intense trauma such as genocide, the annihilation of the Shoah or the ongoing ferocity of the Palestinian Nakba.

To an extent, however, a self-image shaping grounded in the collective imaginary is common to most societies. On a small scale, as in a town or a neighbourhood, imagination and invention pervade local understandings of place even when there is no need or desire to purge traumatic memory. When talking to people about what place *means* to them, listeners are inevitably going to be surprised by the range of responses received (Pink, 2009; Dicks, 2010). Researchers (or simply sympathetic listeners) will be presented with things or interpretations they may have been quite unaware of prior to their questioning. We may not always be able to fathom the full extent of the intersections of the imagined and the real in the way we experience or even understand space and time. This will take us by surprise not because our research has not been sufficiently thorough, but because in this way fieldworkers learn about things that do not necessarily qualify as 'truths' outside of a given co-researcher's imagination. Researchers are successful when they not only gather data and information grounded on fact and observable realities but also learn about the 'what ifs' and make-believes their co-researching communities live by.

"In community, self and terrain are intertwined", writes Suzanne Keller (2003: 11). Often, these ties are imperceptible to the naked eye. Sometimes, perceptions or understandings of the self/terrain dynamics and the world one makes therein are intimate, sacred and not to be shared with strangers (Cappelletto, 2003). This diversity is what, to my mind, makes qualitative research such a precious, intriguing process even when the imaginary edge of some of the tales and interpretations held by our co-researchers veers towards the stuff of legend (see Henare et al., 2007; Black, 2003). The imagination ought to be foregrounded in scholarly research as community-made, community-dreamed and community-performed agency (Henare at al., 2007) across self and terrain.

The logics of co-production enable the building of stories from the ground up, stories that then take on a life of their own as they unfold and reach multiple, ever wider publics. Co-produced fieldwork insights and shared knowledges rework and reframe recent events in unique ways, even if they may appear unorthodox or unwarranted. This 'sharing' approach is especially critical in fields such as cultural heritage, where people's perceptions of the past (and place) are key to their present positioning (Dicks, 2010). How may we convey a version of a group's sense of place which also involves them directly as co-researchers and curators, not simply as participants or subjects? In this book, I attempt to show how we might implement this strategy through collaborative fieldwork and mapping; I also explore the way in which an institution, Beamish Museum in England, is already carrying out work of this kind.

Experiential learning is central to many of the experiments described in this book and is a core theme in my wider research (De Nardi, 2014a, 2016). Rudy Koshar's idea of memory-landscape, for instance, encompasses the material and immaterial elements of memory and landscape as intertwined in experience and encounter (2000). The memory-landscape is said to include objects and markers that we can see, touch and hear, such as monuments, parades, performances and street names, but also wider 'sense of place' (Agnew, 1987), which is arguably closer to a feeling or a sensorial exposition of memory than to a tangible thing. The imagined, then, becomes a quality of affects, atmospheres and things that are tied together by stories and bound together by human and more-than-human experience. "Bodies, things, social formations, ideas, beliefs and memories can all possess capacities to materially affect and be affected" (Fox and Alldred, 2018: 1). It follows that not only are memories materially produced, but that place understandings are created and given meaning through affectual palimpsests of perception and memory. This intersection occupies a meaningful space in social and individual sense of place and identity, and heritage experience is but one of such interlinked affects dynamically negotiated through real and imagined assemblages.

A memory more than most

This book looks at the pervasiveness and endurance of certain memories and attends to the deliberate erasure of others; yet in most examples, presence and absences often coalesce around the same places. The two fight for primacy in the ways that successive generations of people who live somewhere make their world and understand their place therein. Heritage practice is one of many facets of this process of dwelling. But is memory a form of heritage-making, or is heritage a process of memory-making? Both stances may be true. In this volume, I explore their relevance to each other via multiple brecciations of affect, emotion, and place experience. Both heritage and memory depend on the private and public mechanisms of remembrance and deliberate or accidental forgetting. To be sure, memory has become 'big business', especially in the post-war period (Berliner, 2005); this surge in scholarship has generated various strands of research and artistic production attempting to make sense of, and heal, the black hole left in memory by war, conflict (Portelli, 1997; Cappelletto, 2005; Till, 2005) and, in particular, the Holocaust (White, 1990; Young, 1997; Pickering and Keightley, 2012; Kidron, 2012).

The omnipresence of memory is inevitable, as I already alluded to in the introduction; people have been thinking and writing about the process of recollection and remembrance for decades, and this preoccupation has percolated into most studies that concern themselves with the analysis of perceptions of the past in the present. Whether offering meditations on collective memory (*sensu* Halbwachs, 1997; Ricoeur, 2004) or on social memory, or more recently, on activist and marginal group memory, scholars have extensively engaged with the process of remembering and the construction of the past since the beginnings of human history. Others have made a convincing case for the 'gap' in memory (after Bessel, 1996) to be investigated; we have come to appreciate the forceful impact and role of the media in bridging this gap between "the official memory of the state' and memories that are 'current in the private everyday life" (Peitsch, Burdett and Gorrara, 1999: xxiii), versions of the past told by family and friends. Mosaics of anecdotes, palimpsests of recollection, veil upon veil of impressions, imagination and emotion make up the way we relate to the recent and distant memories and positionalities we live in the everyday. Different sets of memory occupy different spaces, geographical and imaginary. In the 2016 US elections, 'alternative facts' were meant to create an alternative set of memories of President Donald Trump's inauguration (Mixter and Henry, 2017). Propaganda-fuelled epistemologies of what is possible or even likely has tinged the social imaginaries of Australians who were made to fear 'African gangs' allegedly roaming the streets of Melbourne (Windle, 2008).

When we come to identify what memory is, we do not stop at the constructs by which societies remember and understand their past, present and future beyond factual events; the past can be known "not through imaginative stories but through the rationalization and the conventionalization of experience" (Misztal, 2003: 118). Imaginary mechanisms that exceed or resist representation are as prominent in a group's perceptual world and its everyday cosmologies. Tall tales, gossip, exaggerations, reimaginings – they all matter, they are all usable data in our work (see Black, 2003; Andrews et al., 2006). Alternative imaginings enrich and complicate the picture (Cole, 2015; Cappelletto, 2003). Objects not only remind us of past practices and situations when connected to memories or associations; objects then act as 'portable places', transporting the self back to distant places and times (Bell, 1997: 821). It is a difficult translation process, and one we all experience in our work. Those of us working with the depth and breadth of memory may often encounter difficulties in establishing an extent of 'historical truth' of 'fact' at the delicate intersection between private, mnemonic processes and collective memories of events (Cappelletto, 2003; De Nardi, 2015; McAtackney, 2016). In the ever more urgent need to decolonise knowledge of the past and representations of past events (Lucero, 2008), however, we have a responsibility and duty of care towards those whose voices have been silenced and whose epistemologies have been negated. In this vein, some scholars question the legitimacy of the production and reproduction of memory within the historical process.

The complicated nature of memory comes across in the expectations for the future that are shared by communities. "In order for our own memory to be aided by others, listening to another's accounts is not sufficient. [. . .] It does not suffice to reconstruct, piece by piece, the image of a past event in order to obtain a memory" (Halbwachs, 1997: 63). The recollection would be too abstract and detached to make sense. Halbwachs goes on to foreground the continuing nature of exchanges and interactions in a group as the main root of pervasive social recollection: the making of shared remembrance. But what about shared remembering as a shared affectual practice? Scholars such as Divya Tolia-Kelly (2010) and Stephen Legg (2005) have identified bias in traditional ideas of collective memory (Nora, 1984; Halbwachs, 1997) as too streamlined towards selective, and usually Eurocentric, 'collectivities' that reflect particular embodied 'Westnocentric' sensibilities and ways of knowing the world (Tolia-Kelly, 2006: 214). Place experience and emplaced memory should instead be conceptualised as raced insofar as they are quintessentially embodied. As professional or avocational social scientists, we are complicit in the shaping and making of memories in our interactions with place and social groups. The idea is that most of the fieldwork we do with our community co-researchers brings up a plethora of impressions, contradictory ideas and memories, affective compulsions and passionate denials that are not easy to qualify – let alone quantify (Rizvi, 2006). The main frustration is born of the simultaneous and contradictory desire to attain and challenge representation. Communities are tired of being spoken for, but will not speak on certain subjects or won't challenge certain deeply held tenets about the very foundations of their existence. At the same time, social groups may feel a sense of impatience, eager to tell their story in a way that exceeds academic knowledge but strives towards a sense of fairness and justice (see also Kiddey, 2017).

The opening up of "histories from below" (Portelli, 1997) has certainly helped gain a more inclusive perspective on the past as experienced by many through erasures and forced silences, as well as through action and resistance. But going back to de Certeau's preoccupation: why make a note, why fixate on some things and not others? In academic scholarship, we tend to draw fundamental distinctions between tiers of historical fact, perception, recollection and understandings (Ricoeur, 2004: 124 ff.). Historians routinely encounter difficulties in establishing 'historical truth' suspended between private memory processes and collective recollection of events (Olick, 2008).

In order to capture and (re)produce multivocal readings of memory landscapes, the reader needs to acknowledge and engage with the multiple purposes of public memory. Without explicitly drawing focus back towards understanding textual or narrative representations of memory as the basis for memory work, it is important nonetheless to acknowledge that material representations of memory are also important; we also need to frame our interpretations through other scholars' readings of memory landscapes. To read representations of memory in place requires interrogating representations of memory; we may ask why a specific event, person or place has been singled out (Drozdzewski, 2015); we need to encase the small scale in its wider frame of power and ideology. In this multiscale realm of existence enacted by memory's tangible and intangible phenomena, we should pay attention to how memory objects can act as custodians and foci of identity and remembrance practices. We also need to understand how and why tangible memory-markers are purposefully designed and positioned to trigger, personify and communicate certain memories over others.

Research at the intersection of memory studies and cultural geography has been a fertile ground for insightful reimaginings of place agencies. So, for instance, Karen Till (2005) has drawn attention to how representations of memory in the landscape – in her case, Berlin – expose both wounded and multi-layered memory landscapes. In places like Berlin, the past cannot fully be erased as long as the remnants of past histories and hidden, buried former identities (Nazi supporters, activists, victims) haunt present-day (re)productions of the city and its memoryscapes. The processes of handing-down memories and places are a responsibility not every society wants to shoulder. And yet someone must situate and work through memories: "past happenings and their meanings are discursively produced, transmitted and mediated" (Radstone, 2005: 137). Transmission and mediation of a remembered past is pervasive and not easily abandoned through time in favour of alternative ways of remembering or other versions of the collective imagination and resistance (Cruikshank, 2006).

We may start from the idea that the interrogation of memory is an embodied, present act that reanimates things of the past through their affectual presence. Where memory falters, the imagined steps in, working through and with memory in the production of identities and affectual tetherings to objects and places. The imagined affects all kinds of past accounts production and rehearsal in the social sphere (Kavanagh, 2000; Witcomb, 2015a). If we accept that a subjective, selective element is intrinsic in the production of all scholarly accounts, *including* written historical documents (and not just oral testimonies), this insight frees us from the traditionally disparaging view of oral history as subjective and biased 'empty talk' (Portelli, 1997; Ricoeur, 2004). That memory could be treated as one of several types of historical source is certainly not new (Thompson, 1998). That memory and its haunting qualities can be treated as a source and manifestation of affectual presence is established also (Navaro-Yashin, 2012). And, as far as 'truth' is concerned, a recognition of "multiple narratives that is characteristic of oral histories ought to protect us [historians] from the self-righteous totalitarian belief that 'science' makes us depositories of unquestionable truths" (Portelli, 1997: 56).

I wish to build upon this architecture of 'memory as presence' by thinking of the act of remembering as a way to engage with, and reach empathy with, a community or social group in order to present, visualise and recall a social past, however the social is constructed and repackaged. As anticipated in the introduction, memory is tethered to place via a myriad mechanisms and processes of affectual engagements that do not happen in isolation. Further, Misztal urges us to solve "tensions between multiple memories" as an "essential concern" (2003: 156). But for whom? I would suggest that the insistence on representing and 'socialising' multiple memories is precisely the concern of this book.

Doing heritage collaboratively

What heritage per se is, or does, has been at the centre of animated debate in the last two decades. A scrutiny of the essence and practice of heritage has developed on a parallel track to what history (as a discipline) represents and what it *does*. Heritage as an act of absence-presencing, in this sense, is the key to affectual heritage experiences. Imagination steps in where memory (direct memory, that is) is unavailable (Witcomb, 2010). As forecast in the introduction, the imagination is perhaps the most important connective tissue linking together the heritage worlds and 'memoryscapes' present in this book (Cruikshank, 2006). I do not use the term 'heritage world' lightly, but rather to indicate the realm which people inhabit, in which they perceive and look after places with a past, linked to these sites in innumerable ways which can be tangible, intangible, made up and hoped for. Crucially, Marcello Canuto and Jason Yaeger (2000) foreground the performative quality and nature of any community – a community, they posit, is made up of elements who interact, whether face to face or remotely, through a shared sense of belonging. The decision to embrace and perpetuate a set of heritage values or beliefs may be quite far from the deliberate act that most archaeologists, historians and anthropologists envisage (Babadzan, 2000: 143).

Heritage has been reinscribed through the tenets of the postcolonial world, its practices dissected under the lens of social justice and collaborative scholarship. Post-modern narratives and the writing of histories as opposed to a singular 'History' – appropriating and authoritative – inform contemporary preoccupations with multivocality and identity across the historical and the social sciences. Thinkers ranging from history to anthropology, from Hayden White (1990) and Dominick LaCapra (2001) to Michel de Certeau (1984, 1984) and Barbara Bender (1993b, 2002, 2006), just to cite a few, share a concern with the fragmentation of historical (and present) reality and its representations. A greater reflexivity in generating historical and heritage narratives posits that past events and their effects on the present are open-ended and that memory is more apt to creating dizziness than providing 'closure'.

In the course of any ethnographic encounter embodied exchanges of memories and impressions take place. Memory has the ability to transform places, people and events through the production of stories (de Certeau, 1984). In the works of interactive, dynamic memory enacting, there is room for multiple stories, multiple versions of the past and resistance to mainstream narratives (Passerini, 1999; McAtackney, 2015, 2018; Sobers, 2017). In fact, de Certeau calls these acts of resistance 'tactics' or 'coping mechanisms' (de Certeau, 1984: xxii); non-hegemonic, marginalised voices thus have the possibility of infiltrating their own innumerable differences, multiple identities and motives into the dominant text (de Certeau, 1984: 41). "The politics of difference in the heritage sphere is [. . .] felt, embodied, intense and dynamically co-constituting the practices of meaning-making and world-sense-making" (Tolia-Kelly, Waterton, & Watson, 2016: 3). In resisting top-down interpretation and exclusionary narratives, we probe into the many ways in which heritage is impacted by social construct and dynamics and how it in turn shapes how social groups position themselves in their landscape. As long as heritage values are clearly and freely voiced by a community their enactment will entail an element of togetherness. Community is thus a way of learning: it enables knowledge about a past that makes sense of, and shapes, the present.

This practice of co-production is rapidly becoming imperative in our scholarly and professional toolkit so much so that we may think about framing heritage production and co-production (together with residents and heritage publics) as an embodied habitus in practitioners' own experience. We may also think of research design and implementation as wholly relational and open-ended, insomuch as the outlining of any research project is dependent on the reactions, interactions and responses of others. Further, we need "discussions about colonial monuments in the postcolony and their varied impacts" to illuminate the ways that some monuments are "repurposed, others appropriated, and still others pulled down" (Rizvi, 2018: 543): thus, both the practices of othering and dwelling operate across social spaces, in the everyday. Whether in the streets, the workplace or the home, the act and ability to 'make oneself at home' involves decisions and limitations that affect the social and operate on private and public levels simultaneously – a doing together for better and for worse: and indeed violence, precariousness and discrimination may be the flipside of intimacy and familiarity (Kiddey, 2014; Pain and Staeheli, 2014; McAtackney, 2018, 2016; Hall, 2005, 1997). In a postcolonial context, the intersections of heritages of 'home' and 'elsewhere' (Tolia-Kelly, 2004a) may unravel in tense, painful ways, but also in constructive paths to resistance (McDavid, 2002): we can build these paths through stories, things and commemorative practices (Atalay, 2006, 2012; Sobers, 2017).

Archaeological and heritage co-production, like engagement, operate "not with specific people or generalized groups, but with the agentive assemblages that create meaning as they define what a project and its engaged people are actually doing in the world together" (Matthews, 2019: 5). Heritage value co-production, then, may be a self-aware and empowering practice in which professionals encourage communities to write themselves and their values in their archaeological, natural and cultural heritage. Co-production practices, framed as affectual and emotional experiences in the heritage process, are a product of the biographical, historical and social time they are interwoven into (Witcomb, 2015a). Co-production logics may allow foregrounding the more-than-scientific side of heritage, enabling the work of enchantment, imagination and dreams (Kavanagh, 2000: 55).

An interest in the very nature of past representation has been with us for some time, growing organically with our fascination with memory; Paul Ricoeur reflected that "from [. . .] shared memory, we pass by degrees to collective memory and in commemorations linked to places consecrated by tradition. It is the occurrence of such experiences that first introduced the notion of sites of memory, prior to the expressions [. . .] that have subsequently become attached to this expression" (Ricoeur, 2004: 149). An archaeology or genealogy of sites of memory serves to illuminate how societies make place and how they orient the present around shared visions of the past. The encasing of memory in place, and, by extension, in archaeological and heritage procedures, is a dynamic process that imbricates people with tangible and intangible essences of identity and culture. The entanglement of 'things' makes memory and heritage work speak to each other in vibrant, dynamic ways through co-production and "the issue of empathy and of the role that objects and places, as transmitters of affect, have in our empathetic engagement with past others" (Byrne, 2013: 606). Heritage as an empathetic engagement may also render heritage a memory-making process in relation to the poetics of co-production: it may thus make sense to situate memory and heritage-making processes within a shared framework of values and understandings; that is when we privilege the social aspect of doing heritage.

"How can objects (understood broadly to include things, images, media and text objects) connect us to others' life worlds?" (Wehner and Sear, 2009: 143). Wehner and Sear pose an extremely poignant question. According to Ricoeur (especially 1996) and Levinas (1998), tethering our experiences to the lived world is a way to communicate and connect, to reach out with other life worlds, past or present, through shared modes of dwelling, being and compassion. But the objects these thinkers advocate tend to be, in the main, tangible things that can be seen, perceived through the senses and engaged with on a basic level in the real world (Ingold, 2007).

A compelling turn to the ethics and politics of the social in heritage and archaeological practice seeks to bridge our sets of epistemologies and creates new perspectives afresh. Engaging with locals' sense of place (Agnew, 1987) opens up the realm of the imaginative as well as the world of the tangible and quantifiable in heritage understandings. Ahmad (2006), in particular, has made a convincing argument for the line between tangible and intangible heritage to be forgotten, or at least sidelined, in favour of a holistic understanding of community priorities that encompass everyday experiences, recent memories and remote knowledge about the past.

Community-facing archaeology and heritage practices excavate much more than physical remains, but forge and reshape sense of place for the communities whose land, territory or city are being brought up, revisited, taken apart and laid out for examination (Atalay, 2006). Framing explorations of the past in their social and cultural context, in their local affectual worlds, enables practitioners to build on 'data' in creating narratives of the past that make sense to local stakeholders and communities. In community archaeology, this striving for inclusion and social justice has been the concern of scholars such as Carol McDavid (2002, 2010), Barbara Voss (2015), Audrey Horning (2013), Colin Breen (with A. Horning, 2017) and Uzma Rizvi (2006) among others. But is all heritage work tangible? What about the act of reaching back or reaching out to imaginary practices, intangible memories of things that are no longer there or intimations of a future not quite 'here' yet (such as the Remaking Beamish village at the time of writing)? Will we experience a collapse of the past in the present and "enter another world" (Gregory and Witcomb, 2007: 265)? What is the lure of things or events that we have not experienced? In this book, I variously consider examples of how the lore of the intangible coheres into place (Ashworth, 1996).

Heritage observes the materiality of the past "to make arguments that are durable, that 'make places', often by privileging a particular narrative, contributing to a particular 'future' of a place" (May, Orange and Penrose, 2012: 3). Co-production is thus a means of circulating ideas of a past experienced in the present by a community and of projecting a community's chosen versions of events towards the future through joint heritage practices. The "manufacture of heritage" (Alsayyad, 2001) is a phenomenon whereby social groups carefully craft a memory narrative which suits particular agendas. The perceived normalising Western-centric vision and agenda embedded in both the UNESCO 1972 World Heritage convention and the 2003 UNESCO Convention for the Safeguarding of Intangible Cultural Heritage, complicit with the logics of imperialist heritage frameworks and ontologies, have been amply critiqued in the literature: this is not the place to add to that rich body of analysis (Logan, 2012; Silberman, 2014; Voss, 2008; Rico, 2008).

In various cases, universally held Western criteria for heritage, when imposed on other cultural frameworks, do more harm than good. This was also the case in Afghanistan and Pakistan during the Taliban insurgency (Flood, 2002). The rise of iconoclasm targeting the Buddhist 'heritage' in this area reveals that "the act of symbolizing – tying certain objects to certain values – sometimes has contradictory effects. It recommends certain objects to the care of those who share these values but attracts the aggression of those who reject them or who feel rejected by them" (Gamboni, 2001: 11). In imposing Western cultural schemata on a Muslim worldview at a time of political turmoil, the prioritisation of heritage spaces over humanitarian aid encapsulated the height of hubris in the insurgents' perception.

Rather, the awareness of heritage's loss of innocence opens up the possibility that far from being a value-free universal value, 'cultural' heritage may represent a set of violences and erasures (Silberman, 2014; see also Tolia-Kelly, 2006; Hall, 2005; Tolia-Kelly and Crang, 2010). We need ever more research that spearheads a drive towards engaging respectfully with the heritage of the silenced and underprivileged (Hall, 2005). After all, whose heritage are we promoting (Atalay, 2007)? Based on whose excavations/investigations? Moreover, for Voss (2018), "Artifacts can evoke powerful responses that draw emotional connections between the present and the past. Yet, responses to evidence of suffering are not automatic. Such emotions are produced within and through what Williams (1977) eloquently termed "structures of feeling" (Williams, 1977 cited in Voss, 2018: 540). Community-led and community-shaped archaeological and heritage practices are as unpredictable as they are powerful – the synergy between locals, activist groups and the 'past specialists' shapes and gives meaning to events, places and things, big and small (Atalay, 2006; Matthews and McDavid, 2012; McAtackney, 2016).

Often, doing heritage fieldwork with communities who have experienced trauma is a life-changing experience. However, participation and involvement can hurt individuals, families, researchers. Trauma itself can become a powerful actor. To give an example, Lynn Meskell has coined the term 'negative heritage' for the memorialisation of histories of political and real violence. For Meskell, negative heritage is "a conflictual site that becomes the repository of negative memory in the collective imaginary" (2002: 558). In warfare and conflict, "intimate and international violences are closely related. Not only are state violence and armed conflict experienced as onslaught in the intimate realm in a range of ways [. . .], but intimate violence is foundational to geopolitical dynamics and force" (Pain and Staeheli, 2014: 4). How these affects and inheritances of violence are inscribed on bodies and monuments is crucial. Trinidad Rico articulates negative heritage as a constellation of locales that "may be interpreted by a group as commemorating conflict, trauma and disaster" (2008: 344) while Christine McCarthy positions as 'incidental heritage' the precipitates of homelessness as a manifestation of materiality (2017).

Therefore, multivocality and inclusion have become paramount to the heritage discourse and its framing as a socially responsible endeavour (e.g. Kiddey, 2014; Waterton, 2014; Witcomb, 2015b). Sather-Wagstaff (2016) has coined the term 'polysense' to capture this very multiplicity of affectual agencies within the heritage experience. Thus, a polysensual approach "centers on the dynamic relationship between the senses, feeling, emotion, cognition, and memory as continually in process" (2016: 18). For activist-archaeologists such as Sonia Atalay, Rachael Kiddey and Carol McDavid, among others, the mission to incorporate and foreground local perspectives in the production of archaeological knowledge is given precedence over the obligation to publish and share data to assuage funding bodies and the academy's regulations on budgeting, publishing formats and timelines and accepted ontologies of practice. Increased visibility and transparency of heritage bodies and processes run parallel to the 'opening up' of self-authorship and decision-making by publics other than academicians and curators. Visions and missions for heritage preservation and interpretation have changed and evolved with time and with political agendas, but before now there had never been a 'boom' in heritage production and consumption.

The risk, for a long time, was to divorce heritage (by definition, 'past') landscapes from the people and communities who lived and worked therein. As a former resident of Venice, Italy, I often marvelled at the superficial and prescriptive nature of heritage tours of the city that would not encourage or even enable visitors to stop to observe the dynamics of street food vendors and gondoliers or to lend a sympathetic ear to the plight of the locals struggling with Wellington boots and walkways on high-tide days. Instead, the (mainly) foreign bodies would be rushed from basilica to piazza, attempting to cram in as much 'history' as possible at any given time through the all-seeing tourist gaze.

The reason I chose the northeast of England and the Italian case studies (for want of a better word) is that they all, to an extent, deal with a trauma and a loss: a loss of identity, a loss of dignity and a memory loss caused by unpleasant events that involved a whole community of human and non-human actors in their wake. They are also, all of them, imagined communities of sorts (in Benedict Anderson's 1983 conception which I go back to in the following); these are social groups who practice a shared idea of the past despite chronological and sometimes spatial distance from the setting of historical events. The maps produced with these communities are assemblages of things, people, places, affects, atmospheres and stories adding up to layered storyscapes. All I did, with hindsight, was ask for spontaneous and unrehearsed expressions of a group's memory as it unfolded. In some cases, if the ethnographic encounter had taken place a week or a month later, the outcome (the practice) of the map-makers would have been completely different. The mood would have changed. The actors would have been different. A stray comment, a stroke of genius or the regret for something said might come into play, colouring the imaginative practice with a different flavour – making memory more, or less, eloquent. Through my participation in the map-making, I have added an unknown quantity to the communities' sense of places. We may even ask what kind of lay geographical knowledges (Crouch, 2010) these encounters generated. Could this perspective expand our understandings of past lives and places operating in the present?

This book presents examples of mechanisms of resistance whereby reclaiming heritage in the grassroots may enable social groups who collectively feel their version of the recent or remote past is overlooked in favour of a dominant narrative to rewrite their past. This is also the case with the mnemonic group of Italian Fascist sympathisers who claimed a mass grave (Bus de La Lum, Chapter 5) as a site of violence against Mussolini's Fascist loyalists instead of a site of commemoration of the Resistance.

Nonetheless, professional amnesia towards locally held values is often subtle. There is increasing emphasis on the value of knowledge held by non-professionals (e.g. Atalay, 2012), as well as criticism by some heritage professionals for seemingly institutionalising both what should be considered heritage at all and influencing the ideological and legal frameworks through which 'heritage' is understood (e.g.; Kiddey, 2017; Ferris and Welch, 2014: 223). It has become clear that, if heritage professionals are to properly acknowledge the avocational experts' input, they have to recognise that often non-professional 'local' co-researchers and experts may come up with a stronger analysis than their own. Local worldviews may trump expert frameworks by injecting more immediately relevant information into the process of heritage-making (Orange and Peters, 2011).

Community co-research: working and remembering with communities

The performance and understanding of memory take place in a social and cultural framework negotiated and shared by a wider mnemonic community of people. Research should enable "the process of knowing through conversations as intersubjective and social. Interviewer and interviewee co-produce knowledge" (Kvale and Brinkmann, 2009: 18). Similarly, we are all familiar with, and many of us work with, the idea of 'community' in our research and practice; yet in the humanities, we do not always analyse how collaboration with community works beyond the scope of our individual projects.

We often leave it to sociologists, social workers and anthropologists to develop and test theories and practices of community interaction which enable and empower non-professionals and the less well-off to contribute to the creation of knowledge. The very meaning and connotations of 'community' is debatable, controversial even, most recently in heritage studies and archaeology. There has been an upswing in community engagement in the heritage sector, mainly in the shape of "informed and imaginative projects that address local needs and local issues" (Newson and Young, 2017: 12; Waterton, 2015). Community-led or community-based archaeology works along similar lines, namely in the handover of partial control of a project to the local community (McDavid, 2010; Rizvi, 2006: 397).

We can expand our outlook to encompass post-structuralist, new materialist and non-representational canons to extend the notion of community to the post-human. Building on the work of Benedict Anderson (1983), Deleuze and Guattari (2004), Bennett (2010) and DeLanda (2006), Harris foregrounds the opportunity to approach and reframe 'community' as not only "made up of humans but also of things, places, animals, plants, houses and monuments" (2014: 77). Here, I would add stories and memories to that list – including thus the intangible elements that influence and even make or break communities. In Wolff's words (2010: 25):

> How does community collaboration come about? What does the phrase collaborative solutions really mean? The answers to these questions have been the driving force behind thirty years of my work with hundreds of communities and organizations. [. . .] In its simplest form, collaborative solutions mean doing together what we cannot do apart.

A doing together what we cannot do apart: an apt way of phrasing the process of shaping and making the maps and visualisation experiments. Community research and collaboration is a creating and a doing together. And in this book, I reflect on some of the ways that groups of people make sense of their shared pasts and meaningful places through collective remembering; this entails a doing together, a working towards an output or multiple outputs that make place and create knowledge. I am most interested in the ways in which this "doing together what we cannot do apart" creates more-than-discursive and non-textual knowledge, which may be channelled through a deep mapping exercise.

A doing together need not exclude individual reflection and 'feeling'. For Keller (2003: 6, my emphasis), community is also about "shared emotional stakes and strong sentimental *attachments* to others who share one's life space". The doing together marries an attachment to somewhere. The realm of the emotional and sentimental does not negate subjective involvement and engagement with places, things, other people and memories. Where is the place of memory in this 'doing together'? Memory, in this sense, is a thing of the body: inner and outer, moving and reaching out to others. It is the story of presents and futures. "Memory is no longer the narrative of external adventures. [. . .] It is itself the spiral movement which, through anecdotes and episodes, brings us back to the almost motionless constellation of potentialities which the narrative retrieves" (Ricoeur, 1996: 114). Halbwachs's (1997) and Nora's (1984) seminal work on collective memory has certainly shaped the ways in which anthropologists, sociologists and historians think about the mechanisms and politics of remembrance. Halbwachs's initial premise is that every social group "develops a memory of its own past that highlights its unique identity" (Misztal, 2003: 51).

Moreover, if history is "the process through which human beings make and remake their lives" (Callinicos, 2004: 2), how does the past of an individual or group 'come through' to the present? Not only does the present open up to our knowledge and access to (a) past, but "the method is opened up to differences in individuals' and groups' visions, auditions, tastes, and olfaction, each under the influence of place, perspective. Position, interests, movement and educated embodied competencies in acting and perceiving" (Ram and Houston, 2015: 10). This is the kind of heritage research and practice where bottom-up social mobilisation and community participation can accomplish or at least foster deeper social cohesion and smoother post-conflict reconstruction (e.g. Olivieri, 2017).

Postmemory: an affectual bridge through time

> Villages appear to dot the landscape haphazardly until an archaeologist excavates the ancient road networks and realizes that all the settlements align perfectly on some ancient causeways, simply separated by the mean day march of the Roman legions. Who has created the settlement there? What force has been exerted? How could Caesar still be acting through the present landscape? Is there some other alien agency endowed with the long-lasting subterranean power to make settlers 'freely choose' the very place it has allotted them?
>
> (Latour, 2005: 44)

Latour poses an interesting and, I would argue, fundamental question to any affectual and more-than-representational reading of heritage and memory: how do we trace the multiple connections of past and present? In this book, I turn my attention to the multiple manifestations of memory as it lives in and across and animates things and places and informs understandings and perception of heritage in the everyday. In articulating the various afterlives of individual or shared materialities of past thing and places, it is useful to triangulate their presence and absence in the world of perception – for things may be absent yet present at the same time depending on how we look at them. This coexistence of presence and absence is manifested through the affectual act of storytelling, of evoking a memory, or in the secrecy and concealment of memories and facts. The absent present haunts a story as it is structured, enacted, retold. Every version of a story or event overrides the one before and the one before that. Absences and presences often compete for their place in remembrance (see Myers and Woodthorpe, 2008); indeed, heritage presence itself may be a form of absence-presencing, a remembering and longing for something other, past and often difficult to articulate.

Heritage can lend presence to absence by drawing on the affectual ties to things that may no longer exist in the perceptual realm, but that still inhabit affectual worlds. Emotional and traumatic memories also constitute cultural understandings, as well as embodied performance (Cashman, 2008). Here we turn once more to Hirsch's idea of postmemory (2008) as a bridge of continuity and closeness between those who 'remember' and people close to them, who are able to vicariously 'experience' events they themselves have not lived through a connection with their loved ones. Further, Hirsh fleshes out the extent of this intergenerational sharing, positing the nature of postmemories as "the experiences of those who came before, experiences that they 'remember' only by means of the stories, images, and behaviors among which they grew up" (2008: 106). As such, affects haunt and shape the journey from memory to its postmemory state – its consequences and its echoes.

Postmemory may be framed as a shared affect tethered to a past event, an idea of the past or a material trace of the past. In the realm of heritage understandings, imaginative engagements with the past constitute so many postmemories of events experienced in place or evocatively imagined (Witcomb, 2010). A postmemory of a troubled event engenders its own special kind of heritage and demands attention through its emplacement and visibility. Often the imperative to make audible the voices of those who suffered has to do with postcolonial projects of reconciliation and social justice. Historian of urban conflict Christine Mady (2018) has argued that, in the context of Beirut, postmemory in a post-conflict urban fabric may be something which can be grown out of through activism. A postmemory is then enacted by conflicting sets of collective socio-spatial practices – memories of wartime division and counter- or postmemories of social struggles try to overcome deeply entrenched divides. Mady draws on Larkin's (2012) reflection on the deep matrix of postmemory in the spatial and the socio-political entanglement of wartime Lebanon. A storytelling of suffering is overlaid with a praxis of activism and hope (see Cashman, 2008; Field, 2008): two concepts, two ontologies not always at odds in a conflict scenario, as I discuss in Chapter 4. In the framework of the post-humanist and non-representational, the urban experience may emerge through affectual encounters or vignettes (after Latham and McCormack, 2009) that draw on echoes of embodied experience and memory to give life to geographies of resistance and belonging (see also Ball-Rokeach, Kim and Matei, 2001). What non-representational theories foreground, for Cameron (2012), is a renewed attention to multivocality and, in particular, to a 'storying' of the world.

This leads us back to stories. The notion of stories, images and behaviours being evoked, shared and reimagined among communities is, unsurprisingly, close to the project of this book. For Sean Field (2014: 125), vicarious memories inherited from storytelling in the family eventually demand to be confronted with one's own perceptions and curiosity. "Over time, the second-generation child translates these affective phenomena into memories mediated by recall, albeit belated" (2014: 125). This parallel understanding and imagining has profound consequences for self-image and one's own autobiographical materiality. Do I carry my grandfather's war burden on my shoulders, or my mother's inherited knowledge? Will these notions and affectual tethers ever be replaced by a wholly 'scholarly' understanding of the European war and deindustrialised northeastern England?

Postmemory can also be an inheritance, however: a gift. Postmemory, being so closely related to kinship or closeness, is central to an affectual heritage. On one level, it would be impractical to attempt to map out and commemorate every crevasse of the remembered landscapes where communities experience their present-absent past. At the same time, the superimposition of a 'map' of memory and postmemory over the spaces of public memorialisation and a comparison of the points of overlap and contrast could enhance our understanding of heritage as a situated and embodied process (Orange, 2015). Moreover, the practices of sharing and imagining heritage values, steeped in affectual intergenerational connections and acts of remembrance, cannot be underestimated. These are themselves a form of postmemory. In Chapter 4 especially I reflect on ways in which the sense of the past experienced in the urban fabric works through a complex imagined and mnemonic frame of reference fuelled by social and political awareness, historic sensibility and emotion. Communities of memory and heritage stakeholders are all potential actors in the sharing and bridging of experiences, stories and memories that create the fabric of place. The echoes of a recent or remote past make themselves felt as communities experience heritage permeating their everyday social fabric, or (as in the case of the anti-memory site in Chapter 5) postmemory can emerge and overwhelm individuals as a powerful uncanny feeling (after Trigg, 2012): someone else's memories become present at a site of atrocities.

Chapter 5 is a powerful example of the haunting of postmemory of conflict. Postmemory, in virtue of being post-, an 'after something' that has happened, is shot through with imaginary tendrils and projected realities which do not, not always at least, permeate or qualify the organic memory of someone who has witnessed an event and can recall it with some clarity. In Hirsch's inception, "the 'post' in 'postmemory' signals more than a temporal delay and more than a location in an aftermath" (2008: 106). In her own words, postmemory is not a movement or an idea, it is the "structure of inter- and trans-generational transmission of traumatic knowledge and experience" (2008: 106). More broadly, it could be argued that postmemory represents the strength of a perceived connection with past events through the sympathetic engagement with loved ones who lived those particular (usually life-changing, traumatic) events.

In Chapter 6, the postmemory of the Grand Electric as a pit village cinema finds its place in the lives of residents too young to have seen it in action. Their rediscovery of the cinema's unknown history becomes one with their own direct childhood memories of games on its deserted grounds. This postmemory of a ghost site and a once vibrant picture-house and that of its decay are superimposed, blending various affectual experiences of the local heritage. Imagination and shared sense of place fill the gaps in remembrance. This synergy is central to the affectual project.

Another connected theme is that of "rememory". In Toni Morrison's *Beloved* (1987), the author forges this powerful notion in order to define the process by which the protagonist, Sethe, recalls moments that have been forgotten but that resurface unexpectedly through intense embodied and emplaced experiences. Rememory also serves to reframe an uncanny kind of 'postmemory' of Sethe's ancestors and African enslaved individuals who lived and died before her time; all these rememories are experienced through what I have referred to as embodied autobiographical materiality (De Nardi, 2016). These affects permeate the individual sphere of memory but also the realm of the social when they become parts of stories told and shared. Do we share knowledge with other generations as a postmemory, or as a tradition, for instance of war and conflict that happened before one's lifetime (Sumartojo and Stevens, 2016)? Or do we use the rememory idea shaped by Toni Morrison and expanded by Divya P. Tolia-Kelly in her study of the far-reaching yet intimate textures and mnemonic practices of female Indian domesticity abroad (2004a)? Both propositions are equally valid if we are mindful of the delicate and often unstable nature of postmemory as an affectual process.

While the present book stems from reflections on the practice of social histories and heritage co-productions (themselves potential acts of rememory), the original context for discussions on postmemory must be noted. In the main, the implications of the 'structure' of postmemory have been espoused in the field of Holocaust studies and memory studies – used to frame the unspeakable horrors of the Shoah. The concept has also been deployed as autoethnographic medium by Palestinian scholar Lila Abu-Lughod to describe her own father's Nakba trauma and return to a ruined village in his homeland after a long exile in the US, interwoven with her own experience of Nakba in the present (2010). I myself have drawn from the concept of postmemory in my own autoethnographic work with Second World War veterans and the legacy of my Resistance-fighter grandfather Memi (De Nardi, 2015, 2016). The intense impact, enduring aftershocks and haunting echoes of war, violence and destruction find a potent and cathartic vehicle for transmission and even educational purchase in the process of postmemory.

However, might there also be a potential for heritage to tap into these fascinating notions? Could postmemories also apply to recollections shared among different generations, which do not stem from acute trauma and loss, but are rather shared 'experiences' of more mundane, less tragic and even positive moments in time? Might we not appropriate the term coined by Hirsh to consider the mechanisms of attachment to places and things, to affectual echoes of things that pass and touch lives, passing through time and places channelled by love and caring?

Whatever the name we choose to designate the act of remembering together – and for one another – through time and place, I propose that these pervasive affects (postmemories or dreams?) might be mapped out as heritage entities in their own right. 'Deep' mapping of social worlds might then become a way to harness the passion and interest of a group of co-researchers. As the academic participants, we could get closer to the materiality and emotionality of past times that resonate in present engagements with heritage (Waterton, 2015). By mapping our own responses, preconceptions and memories in situ, through the textures of today, we can hope to 'feel' and relate a bit more closely to places as we embrace their many contradictions, making peace with their slippery elusiveness and (often) open-endedness. The everyday world of the social is a place where "the unexpected can occur, where change and transition are not only possible but necessary" (Graaffland, 1999: 3). We can learn to live with that and welcome the possibilities this shifting canvas affords.

2 Deep mapping as memory work
Theoretical and methodological implications of heritage mapping

The visualisations and experiments with place showcased in this volume do not just map memory or past times. The variegated temporality of experience means that these outputs map out present perceptions and future intentions. Time is political, as is memory. Byrne suggests that "if people's relations with old things at a local or personal level are complex and illegible (to the state), a key reason is that these relations to a large degree are constituted in the sensory, the emotional and the imaginary, and they are steeped in affect" (Byrne, 2013: 598). Then again, often memories become entangled in stories – and stories, in their multiple and open-ended possibilities and existence, give life to this book's visualisations. It is interesting to reflect on the idea of memory being "illegible to the state" – situated in stories, yet not always visible or public. So in this book I analyse 'good' (and bad), welcome *and* fraught affectual processes as central to the experience of heritage and at the very core of the collective imagination.

The way I see it, heritage is something people 'feel', 'do' and 'make up' as part of their everyday lives. It is inseparable from the everyday and affectual realms in that the many heritage encounters in the everyday are dynamic. A sense of the past grows with experiences big and small in the present. It's a doing together, too; heritage does not mean speaking for somebody, but to let somebody speak about his or her own experiences of the past (and present) whether they are accurate or not. In this respect, the imaginary often trumps the factual. Mapping is yet another way of expressing a group's own knowledges and perceptions. The medium offered by the 'community' deep map or memory map is not an alternative to discourse, but an accessible complement to traditional textual prescription. Formalised texts may alienate segments of the community not familiar with traditional modes of historiography. Sketching, on the other hand, is an act of the everyday, familiar to most; gathering pictures, words and objects may relay a sense of place and an experience which can include multiple perspectives.

The multi-media and multivocality of the mapping process, taking in tangible and intangible 'things' that matter, can open up future projects. We as researchers might then fill the gap between silences and what is "illegible to the state" and the intimate sense of place that individuals and communities nurture. Fieldworkers and communities could collaboratively fill the gap with colourful, approachable and inclusive 'things', objects and cartographic and photographic experiments – who knows what collaborations may lead to? For Wolff (2010: 36), "if we want to create a caring and loving community, then our collaborative efforts must be caring and loving, too". I can only agree with this sentiment. Researchers who care obtain much more satisfactory results not only for the community with whom they work, but also for themselves.

But this is not a social work treatise, either. My readership is likely made up of individuals working or studying in the arts, in the humanities and in museums and heritage institutions, big or small. Taken as a whole, the book is a companion to those working with memory and heritage manifestations that are hard to define and represent within the traditional academic or scholarly canon. As a work on collaborative and open-ended manifestations of sense of place across Europe, this volume opens up more questions about the past than it does answers. What I tried to do in these pages constitutes the antithesis of a definitive work, seeing as the participants and citizen scientists involved in these 'heritage' experiments will have likely changed their minds by the time the book goes to print. Things will have moved on in exciting, unexpected and wholly unpredictable ways.

Fleshing out a methodology

The visualisations in this volume open up and question ideas of home, nostalgia, violence, deindustrialisation, community. The maps/sketches are a way of working imaginatively and openly with communities and citizen experts. These experiments explore ways of linking up and making sense of experiences and interpretations of the deep past and recent history, often on the same map: in this sense, the visualisations invite both professionals and community-based co-researchers to engage with the 'past as present', its meanings and contemporary preoccupations. Computer and internet access, degrees of literacy and formal qualifications become secondary to local knowledge of place and attachment to place in all its temporal and spatial facets. More importantly still, perhaps, taking ownership of a present through their own chosen representations of a past serves to empower communities wherever they happen to be, from the Po plain in Italy and County Durham in England to the Swat River valley.

"What types of genre experiments or styles best facilitate the recounting of truths mutually arrived at, evolving relationships, and shared memories and activities?" (Ram and Houston, 2015: 21). I argue in this chapter that the visualisation experiments we encounter here may just be that kind of genre experiment. Continuing on the theme of giving visibility to the invisible, which I argued for in the previous chapters, this chapter argues that bottom-up, community-led mapping might become a successful tool to chart and 'represent' the ephemeral workings of memory and the imagination in an accessible format and offsets these benefits against a received history of maps, what they are and what they do. Successive sections justify the rationale for adopting this method in the various case studies illustrated in the volume, for clarity of purpose, before letting each chapter tackle them in much greater depth.

Exploring the rationale for applying this methodology in various contexts serves to outline its eclecticism. My work has always been about collating and bringing together different, alternative and disparate understandings and changing perceptions of place and the past; this kind of cartography, conceptualised and drafted by local residents rather than academic specialists, can bring about a celebration of pride of place. There are several ways in which a community can approach this kind of endeavour.

But there is something else; the traditional binary research/practice paradigm is untenable in community-led research. The agency of others and the intra-actions of multiple events make it so that going 'out in the field' with a research agenda in mind, a pre-planned excursus into ethnography, a pre-plugged roadmap of research, are not tenable. Unlike some more interview-centred fieldwork, the practice here, that is, the maps, have shaped the research. The process of creating knowledge has led to the theoretical underpinnings of this book. From the moment the first community in Italy started sketching their movements on a hillside and taking photos with a map in mind, the direction of my doctoral research altered dramatically. The theory traced the folds of the map-in-becoming; the research did not shape the maps but was shaped and led by the maps.

Similarly, the data that populates the maps in this volume, in coming straight from the community co-researchers and other participants, has shaped the research. The output has guided my thinking, not vice versa. I have based all I know and all of my interpretations of what the maps mean, and what worlds these communities inhabit, on each group's desires, visualisations and expectations. Here, I use the framework of the 'make-believe space' (Navaro-Yashin, 2012) to introduce the 'mapping mode' of visualisation of memory and the imagination. Thinking through the 'make-believe' serves to combine the historically correct and the imaginary in the same 'space' of collective belief. My intention in this chapter is to lay the foundations of a methodology which I then go on to explore in greater depth through five case studies in Chapters 3 to 7. In this chapter, I present the motivations and rationale behind the choice of a somewhat unorthodox and established method of visualisation to give presence to apparent absences, to trace the invisible and, ultimately, to 'map the unmappable' in the dwelling environments of social groups and communities. The next sections provide some background information on the genesis of these 'visual' projects and introduce the 'case studies' – or better, the worlds these communities experience and shape for themselves and shared with me.

Genesis

The inspiration to start sketching and mapping came during a warm summer evening in my native northern Italy when I was discussing the outcome of avocational field walking and research with local scholars and archaeologists in the wake of a high-profile publication by the University of Padua; this was a big glossy monograph and catalogue devoted to the archaeological area known as Monte Altare. As we were sitting outdoors, enjoying the evening breeze and sharing a bottle of fine red wine. The predominant mood around the table was one of slight disappointment in the fact that although the professional academic archaeologists in charge of the publication had sought the avocational local scholars' input and know-how, their names would not appear in the book. Their work and support, they said, would not be credited. After some gentle probing (and a couple more glasses), I gathered that the local experts felt as if the 'pros' had taken their site and made it into something else, something dry, quantitative and empty of its inhabitants. Then the idea struck: what if we were to publish an alternative locally told story of the site? What if the locals were to 'play' the bureaucrats and archaeologists at their own game by drafting their own publication, inspired and built on their intimate first-hand knowledges of the locale and its surroundings? A thrill of excitement crossed the table, and the present company's eyes shimmered in the candlelight. Let's do it, everyone agreed.

The idea was that of 'owning' this landscape and claiming our place within it by foregrounding intimate insider knowledge. Little did we know at the time that we were shaping and drawing up an alternative knowledge of the site, like a spatio-temporal slice of the affects engaging and interacting with the local residents. The outcome of this inspired soiree is relayed in Chapter 3, where I 'map out' and articulate our adventures on the hillside of Monte Altare; I start by delineating the many affects and imaginations gradually populating a set of visualisations which would set a precedent in my work. The subsequent research projects I found myself initiating or involved in also became fertile ground for this collaborative, imaginative methodology. But what of the mapping method itself as a tool of inquiry? How to situate my method's collaborative ethos within the toolkit of imperial exploration?

A fraught technique: mapping as making and breaking power relations

Mapping has never been a neutral medium with which to express and channel the world (Wood, 1992: 17 ff.; Harley and Laxton, 2001: 5 ff. See also Crampton, 2001). Traditionally two-dimensional, instructive and appropriating, the mapping endeavour has been complicit in centuries of colonisation and racial violence against the original inhabitants of occupied and 'discovered' new lands (Harley, 1989). In this section, I explore some of the difficulties in accepting 'maps' as a democratic and bottom-up way of knowledge generation and community empowerment. This unease rests on centuries of politically and colonially based cartographic processes which separated, distanced and othered places and people instead of 'bringing them together' (Harley, 1989; Del Casino and Hanna, 2006). In cultural geography, especially, there is still a hesitation to express or display cultural data cartographically: as Perkins (2004: 381) astutely remarks, "theoreticians of the new critical cartography usually employ *words* to extol the virtues of socially informed critiques of mapping, leaving to other people the messy and contingent process of creating maps as visualizations". The politics of map-making thus become a metaphor for our growing discontent, as researchers, about inequalities and ambiguous positionalities (Crampton, 2001; Del Casino and Hanna, 2006). This ontological anxiety can affect researchers in undertaking or even designing most forms of participatory and immersive fieldwork and engagement with communities. Seldom has such discontent been analysed and unpacked as in the case of mapping in cultural-geographical fieldwork and research (Pinder, 2005).

Humans have long been driven to represent, visually, a variety of experiences and intents. Yet for long the objectifying eye of the Enlightenment cartographer has tended to take sole charge of what was in earlier times a non-appropriating, non-essentialising way of being in the world (de Certeau, 1984; see also Ingold, 2000). The connection between journeys and map-making had become lost in the logics of Renaissance and later Enlightenment modes of spatialisation, with spatial units of measurement overtaking and eventually replacing temporal and bodily parameters such as days of travel, paces and other embodied and human-grounded parameters of spatial movement (de Certeau, 1984). Therefore, a common critique against maps is that they distanced, abstracted and separated the map-maker (and user) from the geographies and materialities making up the real world one wants to represent. There are notable exceptions, when mapping becomes a critical and radical tool for catharsis (Blum and Secor, 2014), vindication and social justice (Del Casino and Hanna, 2006). Indeed, maps, like other imperialist tools, can be upended and owned by subjugated and silenced subjects; map-like visualisations can become part of the toolbox of resistance and decolonisation (Brown, 2015; see also Byrd, 2011; Sium and Ritskes, 2013; Callinicos, 2004). They can be claimed as one's own.

Despite its various conceptual drawbacks, the map has been persistently deployed to represent spatial distribution, be it qualitative or quantitative, of cultural values and social and political processes (e.g. Zerubavel, 2003, Tolia-Kelly, 2004b, Massey, 1995; MacKian, 2004). This 'appropriation' in itself points to the relevance and potency of the map as a vehicle or vessel for cultural meaning; in other words, the map is a pictorial or visual conveyor of storytelling, with the latter being a much-loved trope in cultural geography and more-than-representational theories across the social sciences (see mainly Cameron, 2012). There is more to map-making than the traditional quantitative 'distribution map' canon commonly used by heritage professionals, landscape archaeologists and historians (Perkins, 2004). The autobiographical character of map generation increasingly matters. "There is a reflexivity that is concerned with the role of subjectivity and the researcher in the dynamics of knowledge production as this occurs in the time and space of fieldwork. This is known as research positioning" (Castañeda, 2008: 41).

Similar concerns have increasingly weighed on the role of cartographer or map-maker (Harley, 1992). The act of visualising spatial relationships or spatial understandings of research has incrementally acquired an ethical and moral dimension of reciprocity and transparency of representation. In Figure 2.1 some of my second-year students at Western Sydney University spontaneously sketched a perceptual map during my lecture on community heritage mapping. They conveyed

Figure 2.1 A 'visualisation' spontaneously made by some students in class during a workshop on community memory mapping at Western Sydney University

Source: Reproduced with permission from students

the main topics and ideas as a storyboard, in which I am depicted smiling benignly. It made me very happy. Mapping can be done anywhere and can enrich or overlay multiple things or meanings to an event – in this case, a traditional classroom learning experience.

The push towards 'better maps' has continued to grow. Jeremy Crampton (2001) provides an account of the development and new ontologies of mapping in geographical fieldwork that can be applied to other social sciences and the humanities. His book represents an epistemological turning point in the practice of mapping and is addressed in particular to the (then) emerging GIS-based cartographic practice. Crampton (2001, 2010) was the first to critically acknowledge his own positionality and presence as geographer and map-maker in the field. He critiqued two-dimensional ways of mapping, identifying and registering the synergy of temporality and spatial sense in the experience of learning about place. This work has been paramount to the thinking that informs this book and my practice, above and beyond its use as a community knowledge co-production tool.

Traditionally, the field experiences of ethnographers and cartographers could not be more different. The former often encounter difficulties in establishing 'a final version', aware of the pernicious process of interpretation but also of the need to identify a narrative they can work with and publish. Cartographers, on the other hand, have habitually observed space, calculated, measured and reproduced its dimensions and configurations. Ethnographers question the very nature of the production and reproduction of memory within the historical process. Most cartographers simply used to 'get on with it', going through the motions of map-making in a confident habitus of spatial knowledge production (*sensu* Bourdieu, 1977). Things are changing rapidly, with disciplines and fieldwork logics cross-pollinating; this book, then, hopefully consolidates the notion that being in the field and making maps are in fact not mutually exclusive and that these two practices speak to each other through the complexity and togetherness of human experience (Pinder, 2005). The lone cartographer scribbling away on paper in a closed room away from the world and people therein is a thing of the past. Parameters have changed. Mapping is an embodied act, a performance of the body's orientation in place and space, even if has been traditionally conceptualised as separate from the senses and perception. The experiential maps of today are made on the fly, while present in the entangled geographies of the right now, among others, human and non-human agents, all contributing to one's sense of place, the loss of one's bearings, the finding of one's way. Body, or bodies, stand out as emplaced vehicles of affect and emotion, sites of agency, of endlessly unfurling possibilities, and emerge as subjects rather than Cartesian 'second-class citizens' subjugated to the objectivity and primacy of the mind's eye (or bird's eye view). Whatever the motivation for 'mapping' experiences, the reliance on the primacy of sight is no longer a given (Pink, 2009). Contemporary and 'new' map-makers are opening their methodology to the complex multi-stimulation of the perceived world (Thomas and Ross, 2013). Dynamism and collaboration have entered the cartographer's toolkit.

The groundedness of maps in the present

The map's implications in colonial acts of genocide and invasion cannot be disregarded in any discussion of its viability as a tool for social good (Crampton, 2001). And yet, as we have seen, map-making is shedding its murky past valence as instrument of oppression and progressively opening up democratic possibilities and strategies for self-determination among communities (see Wood, 2002 for a reflection on Brian Harley's new manifesto for 'open' and dialogic maps). To disrupt the map is to move into a terra incognita of methods, projects and desire which resist hegemony and tradition; it is a breaking through into new ways of expressing place and place attachment.

Memory work with people and memory objects is a step towards collapsing 'the experience/analysis divide, such that the experience of things in the field is already an encounter [. . .] with meanings" (Henare et al., 2007: 4). Where oral history and archaeological practices on the one hand, and memory-visualisations and maps on the other, part company is the point where we conceptualise their relevance and justify their existence. The oral historical exercise extols and probes memory of a past; the archaeological investigation discloses through destruction; the maps create and build on a present through recent and urgent understandings, plans and imaginaries that live in the present moment. "There is a concomitant difference between the methods of ethnography, which primarily focus on *the present day*, and lived/live information, and the inherited (or traditional) archaeological methods, which primarily focus on *the past* and its materially accessible manifestations" (Castañeda, 2008: 36, emphasis in original). Yes, so the maps do visualise things which have been or in some case have ceased to exist, but they fundamentally embody and channel present emotions, made in groups of living individuals who get together for a purpose. This, I contend, may be the main merit of this methodology.

The very recent publication of a map outlining the sites of the massacre of Indigenous Australians by colonists/settlers seems to stress the fundamental idea that even when a politics of memory strenuously and doggedly attempts to remove certain narratives from the landscape, the memory traces remain (Lyndall et al., 2017). These traces of invisible deeds persist – in local narratives, in the embodiment of those places, in the (hidden) archival documents (Butler, 2009). It is crucial to this book to emphasise the importance of map-like activist research as in the Australian example prior and move beyond the strategic othering and distancing exclusionary practice of much former cartography. The Australian initiative shows that two-dimensional mapping can still express the living and dying experiences of subjugated individuals and communities in a traditional form (a distribution map) and through the open-source, high-tech medium of digital mapping.

I now briefly review the mapping projects in this volume and start from the mapping process to gauge why each group of co-researchers chose to tackle such an endeavour. The book first examines the communities who live near the 'heritage site' of Monte Altare.

Monte Altare: when we put 'ourselves' on the archaeological map

During my fieldwork with them as early as 2003, I had become aware of a discontent, a dissatisfaction with the way the local non-professional historians, archaeologists and sympathisers had been shut out of any and all academic publications on the site. In a way, the map-making exercise at Monte Altare became a playful rebellion, a one-off act of irreverent defiance aimed at the professional architects of the archaeological excavations, artefact collection and official study of the site. Enmeshed in an affectual geography of home, Monte Altare is instead filtered through by many moods and understandings that blend with archaeological and antiquarian knowledges in a holistic topophilia, a place attachment relevant in the present as more-than-a-relic.

In fact, the Monte Altare community map-makers were keen to encompass present and past attitudes to the site and to incorporate the fear of medieval dwellers alongside the affection of contemporary residents. The artefacts and figurines stemming from the site, lovingly and painstakingly collected and looked after by local avocational scholars, become visible once more on the map alongside photographs and comments by participants. These objects had been hidden, as were the insights and emotional stakes of the community magically erased by the official academic publications and sterile catalogue that were issued on Monte Altare, listing objects extrapolated from their affective framework. "We become alienated – out of line with an affective community – when we do not experience pleasure from proximity to objects that are already attributed as being good. The gap between the affective value of an object and how we experience an object can involve a range of affects, which are directed by the modes of explanation we offer to fill this gap" (Ahmed, 2010: 37). Publication and separation of material culture from the society that found it and nurtured it is an act of severance leading to a dissonance, to an othering. The once-loved field finds had become unrelated miscellaneous items in a catalogue. Extricating artefacts and archaeological objects from their affectual context, their web of local identities and entanglements, despoils them of their value.

On the maps we made, local understandings came once more to the fore as rich engagements of local place names, hearsay and feeling bodies in situ, and the maps shaped experience and fieldwork as one and the same. These memories and their telling were mediated not only by the body, but also by the cultural context in which they were produced and enacted. Heritage-related fieldwork does not merely produce narratives embedded in either the interview subject or the researcher, but, rather, "the process of knowing through conversations is intersubjective and social" as interviewer and interviewee co-produce knowledge (Kvale and Brinkmann, 2009: 18). The knowledge engendered and captured in these make-shift maps could be thought of as popular knowledge as well as academic knowledge for two reasons: these impromptu and non-hierarchical perceptions define cultural belonging and sense of identity of the interviewee and of the community to which the person belongs; second, this visual narration, or rather, performance of memory, illuminates heritage practices' place in a social framework negotiated and shared by a wider mnemonic community of people.

The political message of creating and sharing alternative, locally made maps of a heritage landscape is perhaps subtle, but it still drives a message home: heritage landscapes are *always* someone's landscape of home, someone's local place (see also Orange, 2010). The disruption of abstract and top-down map-making at a site like Monte Altare serves the purpose of opening up potentialities for further similar engagements by other communities, as I argue in a paper describing the methodology (2014a). In other words, the maps which the co-researchers and I made on the go and later perfected reflect the topographies we negotiated, while not neglecting to record and contextualise meanings. We mapped physical place while mindful of the links between fictions, histories and 'sense of place' encountered along the way. Far from being parenthetical, these ideas and imaginings became protagonists on the maps, as we perceived them with an intensity impossible to muster for a distribution map. The resulting maps effectively trace the life cycle of meaningful locales as well as defining the actual and imagined topography of places.

The Italian war experience: a town, divided

The visualisation exercises in Chapter 4 resonate in different ways according to their users and audiences: the map experiments will stir a spectrum of responses based on the knowledge and the experiences of users, mapping participants and individual political positioning. The reasons for this are many, but first we need to contextualise the places where the chapter's actions unfold. Spatially and culturally, the Italian northeast is a borderland. It feels like a world between two worlds: a land perched between continental Europe to the north and the Mediterranean peninsula to the south. The northeast is a border territory in which different ethnic and religious groups live and where multiple episodes of foreign occupation and civil unrest have shaped a unique cultural landscape that is rich in contradictions. In the decades preceding and following the Second World War, ideas of nation and ethnicity have shaped long-lasting racisms between ethnic Italians and Yugoslavs across the eastern border of Italy (see Hrobat-Virloget et al., 2016). That same coastal strip and the hinterland stretching up to the northeastern Alps were also annexed by the German Reich during the years 1943–1945 when German-speaking, Berlin-administered Nazis ruled the land, overriding the power (increasingly belittled and declining) of the home-grown Italian Fascists.

Mapping a movement like the anti-Fascist and anti-Nazi Resistance and its concomitant civil war in such a geopolitically fraught context has advantages and disadvantages. The Resistance (or liberation war) is enmeshed in everyday Italian consciousness and culture, whether implicitly or explicitly, via an omnipresent – yet ambiguous – collective remembrance process, and dealing with the civil war or Resistance entails a highly localised process of remembering. Official and unofficial memories and storytelling practices of the eventful '20 months' of the Resistance permeate the fabric of virtually every city and town in northern and central Italy. Wartime stories are promoted and facilitated by the outreach effort of dedicated associations – either by gatekeepers or, less and less due to their age, the veterans and primary custodians of war memory. Typical outreach action takes the form of media initiatives, publication and education. Luisa Passerini (1999) and Philip Morgan (2009) argue for a multiplicity of memory levels and meanings, positing that the state-constructed 'national Resistance memory' can and does jar against the largely hyperlocal nature of the Italian experience.

Today, living in an affectual and spiritual borderland in the northeast of Italy signifies inhabiting a territory and an imaginary shaped by bitter and bloody infighting between Italian Fascists and Italian anti-Fascists (of various political colours and leanings). Divided cities also became border cities separating 'us' and 'them' – an affectual economy of alterity and exclusion via the politics of ideological difference. Despite the fracturing of emotional, affectual and socio-political perceptions and experiences among communities in the modern-day Veneto and Friuli-Venezia Giulia regions (much as most of northern Italy, it must be noted), and regardless of the trauma of separation and mutual dehumanisation that divided said communities, the post-war Italian government decided to put the most obvious and unsavoury aspects of the war to bed in a bid to move forward as a democratic nation. Italy was not alone in this selective self-styling: for almost 70 years now, European resistance movements have been perceived and understood as the grand narrative engendering and legitimising the rise and flourish of European democracies after the chilling Nazi parenthesis (Cappelletto, 2005).

After the war, the leaders of the Resistance movement felt obliged to present a united front to Allied victors in order to give Italy a chance to move forward as a democracy. The more insalubrious aspects of the civil unrest were expunged from official discourses on the Resistance while other, less politicised narratives were favoured: this 'filtering' process led to the ad-hoc creation of what could be called a 'best case scenario' of Resistance, a narrative fraught with patriotic motives and values – in the forefront, loyalty and gratitude to the Anglo-Americans after the Armistice of 8 September 1943. "They [the Communists or Partito Comunista Italiano] underplayed the fact that the Resistance was a fight. They underplayed the fact that the partisans were armed fighters who had not only died, but had also killed; all the monuments to the partisans are monuments to dying partisans", writes Portelli (2003: 38). Following Renshaw, we may engage with the dead and their traces as "part of the material culture of the living, used to make representations about the past, the present and the future" (2011: 27).

The uncomfortable questions of 'who the enemies' were and 'who fought whom' haunts the present (Pavone, 1991). Remembrance narratives of the civil war convey a sense of nightmarish confusion perceived by many as the difficulty of navigating and dwelling in places, landscapes, towns, neighbourhoods where suddenly danger lurked everywhere (Behan, 2009). This is a grave act of neglect, seeing as the perception of enemies, of place and of identity has everything to do with the development of resistance movements and with the establishment of regimes. And what about the unlikely, the uncanny, the unexpected stories and memories inhabiting and circulating certain affectual mementoes of the conflict (De Nardi, 2014c)? What room is there for those kinds of feelings and sentiments or even sensations that cannot be represented (Curti, 2008)? Can we use these affectual entanglements as data?

Place, interrupted

What did the community researchers want? No one staked a claim to closeness or personal investment in the place. It was clear that Bus de la Lum was perceived as a non-place.

Some of the co-researchers wanted nothing to do with the idea of a map. Some wanted to put the many ugly legends of partisan atrocities to bed. Others were keen to match a face to some of the victims and to name the Fascist [and collaborator] dead, as if evoking a name could disperse the spell of forgetfulness.

Engaging with experiences around a mass grave has been the single most difficult time during fieldwork; my field diary entry on day 1 begins with an anxious question: what did the community researchers want? I had at that point wondered whether mapping a black memory would do more harm than good. . . . In Chapter 5 I report on the process of working through memories of a place where, at various time in the recent past, Resistance fighters have inhabited frightful agencies as mass murderers or more compassionate roles as the reluctant perpetrators of political executions. The fact remains that Resistance fighters hurriedly concealed the bodies of political enemies and civilians who acted as spies to the Germans, after trials which can be considered summary at best. As we get closer to these fraught understandings and *mis*understandings, I trace the genesis of the visualisations of the 'anti-memorial' site of Bus de la Lum. In the midst of these uncanny anti-geographies of the war, some locations took on a mythical aura or atmosphere: imbued with savage tale of violence and revenge, the sinkhole of Bus de la Lum visited in Chapter 5 is an exemplary case study for the workings of counter-memory, denial and fractured remembrance. Described by a co-researcher as a 'failure' by the Resistance, the mapping of this place reflects negatives that so powerfully outnumber the 'positives' in our visualisation and memorialisation endeavour. The grassroots feeling was that the sinkhole represents a missed opportunity, or indeed a failure, by the local resistance in the region to stand for the nobility and bravery at its source: the struggle to rid Italians of Fascism and Nazism.

For Beverley Butler, "apprehending memory-in-conflict and the 'right to remembered presence' requires further archival modelling and alternative modes of representation" (2009: 68). Perhaps the more-than-mnemonic maps of the civil war may go some way towards Butler's proposition. New categories or extraordinary categories of knowledges and spatial experiences call for new methods. The section of this book devoted to the Italian war experience foregrounds the practices and experiences, "new categories of person or types of object, from a series of intersecting processes, entities, and forces in ways that might be unexpected, unforeseen, and unplanned" (Fowler, 2013: 27). One of the neglected areas of European conflict studies has been an investigation of the processes of personhood and interpersonal perception of actors in the Second World War beyond broad categories of 'friend' and 'foe'. What is still missing from the current attention to multivocality in a war and conflict context (but see Newson and Young, 2017; Olivieri, 2017) is a truly critical focus on the nuanced experience of multiple selfhoods and positionalities in the same place – overlapping, fighting for attention. Mapping identities together with places might go some way towards bridging the gaps in the current understanding of European war experiences – experiences in the grassroots.

An Italian town, the focus of Chapter 4, is a case in point. Vittorio Veneto is split down the middle, its imagined historical geography a borderland split down the political and ideological divide in relation to the Fascist experience. As a border town, Vittorio Veneto has its black spots and no-go areas. Thus, Vittorio Veneto teacher and historian Lajolo wrote,

The reconstruction of events contextualized in the local area and within the wider scope of national history makes students aware that "the great wheel of history passes through here too", in the provincial town, in the suburbs, and that it involves common people.

(Layolo, 1998: 240)

Through ethnographic work with veteran associations and local historical societies, I have explored the cultural and affectual cartographies of the city and its territory, trying to fathom the points of breakage of the life of a community once relatively peaceful. We uncovered together a cultural geography that was duplicitous and often painful to explore – where buildings had double lives and schools and castles became torture chambers. Superimposing spectral geographies to the everyday places of shopping, learning and work served to open up concealed place memories and to attune sensory understandings of a difficult and often murky past. More on this later.

The northeast of England: dwelling among bittersweet memories

We could also consider the framing of the coal mines in the English northeast as a *locus amoenus*, as a spatio-temporal bubble inside which the golden age of the busy economic and cultural lives of communities thrived. As I was sitting in a hairdresser's shop in Langley Park, north Durham, on a chilly May morning in 2016, Linda, the lovely stylist who looked after me, started talking about her family. Her father had been a pit man (a coal miner) since the age of 12. He had been working down the local Langley Park pit until he had a nasty accident at age 45. Then he had been moved (forcibly) from his miner's accommodation with his young family and sent to another pit, where he would look after the ponies as he was too disabled to do a 'proper' pit man's work. . . . Linda told the story in sober tones, but I could sense an undertone of melancholy sympathy. My heart filled with an infinite sadness at the thought of this man, Linda's father, who had ended up doing menial work for meagre pay after being pushed out of his regular job by a bad injury. How unfair and how cruel that world had been. Linda, and a colleague whose name I did not catch, affirmed that every pit village had its history of tragedy: the mutilations, the injuries, the subsequent disengagement and isolation of the disabled miners. The memory of the accidents lingered in the local topography like a resigned, almost hagiographic geography of wounds. Now, these same communities in the post-industrial British regional areas (northeast England, Wales, Scotland) find themselves deprived of that energy and former livelihood (Walkerdine, 2010; Smith and Campbell, 2011): the landscape itself, laden with pollutants in the soil, manifests itself as a melancholy and even bleak afterthought of industry (Maantay, 2013). After sustained neglect on the part of various British governments consigned entire pit villages in England and Wales to unemployment, poverty and, to an extent, social marginalisation after the closure of the mines (Fieldhouse and Hollywood, 1999), a pervasive idea of the golden age of the coal mines exists; this nostalgia coexists with the awareness of painful memories of harsh labour exploitation by which thousands of boys and men died and were injured. The collective imagination may trump individual stories of suffering or live alongside them; this entanglement depends on *whom* one talks to.

And yet, good or bad, that world was gone, and the whole region, the whole community, was still mourning for it (see Benyon, Hudson and Sadler, 1986). The London-centric establishment and subsequent governments had perpetrated an irreversible act of violence against the people of the northeast by closing down smaller local coalfields, aggregating pits as 'Super Mines' and thus tearing apart their pit villages and livelihoods without a thought about their quiet, laborious dignity and hard labour (see also Strangleman, 2001). There was fierce pride and a sense of social worth in the words of many of the locals I came in contact with in County Durham, Sunderland and Gateshead. I could connect Linda's story to many other memories I had encountered since moving to the northeast of England. Black and white photographs showed smiling men, women and children, proudly posing underneath the local coal pit's banner. In other stories, working men's clubs and schools in pit villages were full of life, bustling with activity. Visiting these places now, these long-forgotten pit villages mourning for their vibrant former life, one does feel a sense of a past betrayed – a lingering memory of disappointment and hurt. I thought of these assemblages of people, places, things and animals once working organically as a community: the pit ponies had names and were looked after by children and elderly men or those (like Linda's father) who had become too ill to work in the pit itself.

I remember a moving picture of a man looking down at pony called Kent with quiet affection in a photograph dated 1962. In another, haunting shot, the blackened face of a miner in the background stared out at the viewer with an enigmatic look, almost mask-like, juxtaposed to a (relatively) clean-faced lad stroking a pit pony in the foreground. The slow decline and closure of coal mines in the English northeast in the post-war period is framed through the lens of the imagination, intangible perceptions and affective biases. The remembrance of the coal mining industry foregrounds emplaced stories and values through collective knowledge building and exchange (see also Orange, 2015).

In spatial terms the hauntings affect the landscape itself, not just prints in the social fabric of life. The ghostly shapes of the colliery *head gears* still haunt the semi-rural landscape of County Durham, punctuating the landscape from Geordie to Mackem lands (Newcastle-upon-Tyne to Sunderland) like accusatory fingers. These silent remnants of industry stick out of the soil like lonely ghosts of a past lifetime. The land has been populated with legends of a golden age of mining industries and prosperity, although the historical imagination in these parts has somewhat embellished the historical truth of the times. The spatial configuration of the memory of the coal mines in their wider landscape is mirrored by village-specific remembrance practices that came to life on the maps presented in Chapter 6: different scales of memory and rememory, beyond nostalgia.

Surviving or not

What materiality did the coal mines entail? The pits' decline and closure has left many a nostalgic void behind it, but at what cost? Living and working at the coalface gave no one an easy ride: there was a high risk of death by accident and almost inevitable long-term damage to the miners' health. The women in the community, and the pitmen's children, themselves worked for hours, tending kitchen gardens, cleaning, raising infants. Wives waited at home in solitude for up to 17 hours a day. At the end of a pitman's career, things did not get any easier. Many aged miners ended up in care homes where charred, blackened lungs kept each other company. The so-called black lung disease often marked the bleak end of the men's lives if they had not been poisoned by mercury or crushed in the poorly ventilated mine shafts (Meiklejohn, 1952).

The photographs and stories of coal mine lives act as a kind of stubborn industrial memento mori: they are present here to remind you that life was precarious, but it was still a life worth living. Kidron has reflected on the nature of memento moris, on objects that reflect an unnatural or premature death, accident or loss of a person or persons, as powerful agents of mourning and myth-making for communities and individuals. "Souvenirs of deathworlds retain a no less subtle balance between the representation of difficult memories of rupture from loved ones" (Kidron 202: ii). Thus, objects that silently encapsulate and perform pasts that have culminated in death or near-death experiences are particularly complex. Within the family living room display, objects are caught up in the materiality of their setting, able to move between past and present or channel affects to those who view or handle them (Hirsch, 1996). The point is that the memory world of the mines is neither a deathworld nor a past world. Some of these objects "do not evoke natural or anticipated death but, rather, dramatic and unexpected ruptures in the texture of the self and the family" (Kidron, 2012: i). But then again, these objects are almost meaningless without stories. Seen without context the lamps and tools that belonged to the dead or crippled pitmen would not move us to tears. "What are the stories we like to hear? [. . .] they are often the ones that confirm us in comfortable ways of thinking" (Turkle, 2007: 323).

"The end of the story", Ricoeur writes, "equates the present with the past, the actual with the potential. The hero *is* what he *was*" (1991: 114, emphasis in original). The coal mining heroes are such because of what they did, what they represented, and not necessarily due to their individual identity. Their respective pit village communities are all caught up in the pitmen's fate, enmeshed in their accomplishments and entangled in their legacy. Memory can transform places, people and events through the subjective production of stories (de Certeau, 1984); memory is relational – never occurring or developing in isolation, but along with relationships – it is "a sense of the other" (de Certeau, 1984: 87). The survivors in families looked after each other until they were moved on to another place. Since the mining accident victims cannot tell their own stories, the responsibility falls on the community to make sure they are heard. It falls on the miners' loved ones to share their legacy. How the community chooses to construct the past it shares with the dead depends on several factors that include the intended audience. The musealisation of the loving heritage of an industry and its heroes has contributed to how these mnemonic narratives are shaped (more of this in Chapter 6). Some themes in said narratives include the affective interplay between history, memory and place; an emphasis on the dignity of work; the importance of social support and of integrating the social and the cultural against displacement; the incorporation of meaningful silences into the verbalised stories.

Doing research with these villagers, past and present, we begin to explore the nature of coping mechanisms, and we can lift the curtain on the emotional dimension of disenchantment with the state, usually an issue explored through the prism of political activism and social unrest. Moreover, by looking beyond the literal, factual and 'naturalistic' interpretation of stories about the pitmen as community heroes, we can delve deeper into their role and place in wider affectual histories (and memories) of a community. This is not what the mapping experiments in Ryhope and Kibblesworth aimed to do: rather, these community research and knowledge production projects incorporated the memory of miners within a wider affective economy of place.

It is now time to delve deeper in the stories and the experiences of these various communities. In the next chapter we begin our review of methodologies and interpretations of past, present and future starting from an open-air archaeological area in the northeast of Italy – the first stop on a journey to playful, imaginative place-making practices and visualisations.

3   Imaginative engagement

The sleeping giant and the cursed hill

The hills have eyes: the legend of Monte Altare

The spring breeze accompanies us on our journey up the verdant slopes of Monte Altare. The hill, some say, has experienced, tolerated, facilitated and welcomed human habitation and use since the Bronze Age or even earlier. Whether this intense and intimate relationship of the hill with human beings has always been peaceful and consensual is another matter. As we climb the slopes, pausing every now and again to regain our breath, take photographs or take a drink of water and share comments on our experience, we are here now – this is our time on the hill. The locals know that this place has always been important: we are but a chapter of its long, turbulent but ultimately loving and caring history. The hill has agency insofar as it has shaped dreams, imaginaries and attitudes of local dwellers through time.

Introduction

This chapter tells the story of a group of volunteer archaeologists, community-based experts, avocational historians and interested residents engaging with the past and present of a particular landscape in the Italian northeastern province of the Veneto. The stories within develop as locals encounter and engage with their 'heritage' landscape – more precisely, the environs and slopes of a hill known as the Monte Altare. What follows is my attempt to record, sketch and report on the many meanings and impressions – often unorthodox and subjective – we conjured up through our collaborative fieldwork. In this chapter I relay the imaginative workings of an imaginative and playful social-cultural geography in practice as they were then 'displayed' on a series of heritage maps. The explorations and musings on place and time within these maps stem from the point of view of a local community informally involved in heritage and leisure-related practices. This community encompasses active members of local non-professional Archaeo-clubs (Italian archaeological societies) and, through the latter's promotion of their work at a local level, the wider community attending open events and activities.

The initial aim of the mapping exercise was to trace contemporary heritage understandings from the bottom up, which would then form the basis of my

Figure 3.1 The Monte Altare
Source: Photograph by Sarah De Nardi

doctoral thesis, alongside 15 more Italian case studies (see Figures 3.2 and 3.3 for two examples). With this in mind, I also discuss issues of power balance and representation that I encountered in the field: mainly, I highlight the oft-perceived lack of equal access to, and interpretation and understanding of, heritage places from the perspective of residents acquainted with their heritage landscapes, in opposition to 'outsiders' (bureaucrats, institutions, academic 'experts') treating the area as the subject of academic study. Representation of the local heritage sites around the Monte Altare in mainstream archaeological outputs seems to have occurred without the establishment of a significant dialogue with local groups. This neglect often exasperates local communities, and in the course of this particular fieldwork I identified a discontent and a desire to be heard, to have a say in the management, presentation and interpretation of their past which agrees with current preoccupations, moods and expectations.

In a nutshell, my fieldwork between 2002 and 2008 contextualised heritage activities within the leisure time of a community of people in northeast Italy. In the locals' perception, the chronological details and use and settlement phases of a heritage landscape (which are the archaeologist's bread-and-butter) made no sense if not as part of a bigger story. Time did not make sense to them *on its own*: the community experienced the site as a living part of the fabric of their everyday lives (see also Waterton, 2005). And this place-making process existed in place, developed and shared collectively, without necessarily having to do directly with any tangible – or visible – forms of archaeological heritage: material culture.

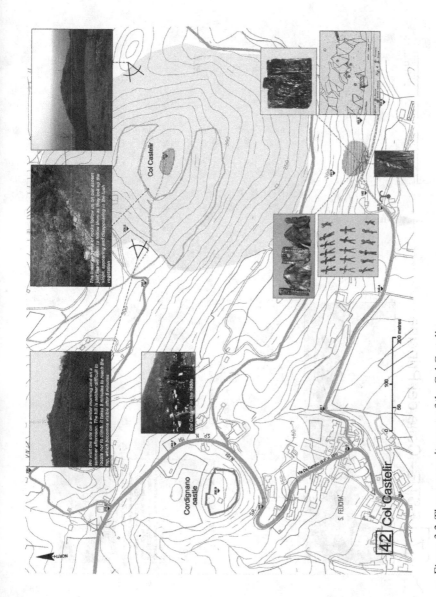

Figure 3.2 The community map of the Col Castelir

Source: Copyright Sarah De Nardi

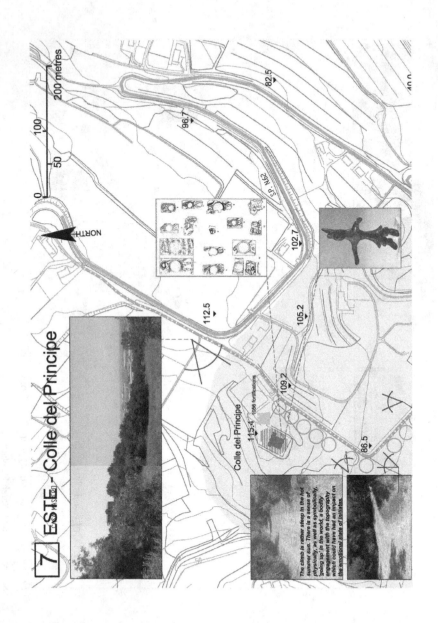

7 | ESTE MUSEO – Colle del Principe

NORTH

0 50 100 200 metres

S.P. N.62

112.5

96.7

82.5

102.7

105.2

109.2

86.5

115.4 1066 fortifications

Colle del Principe

The climb is rather steep in the hot summer sun. There is a sense of physicality, as well as symbolically, being up in the world, a bodily engagement with the topography which could have had an impact on the emotional state of initiates.

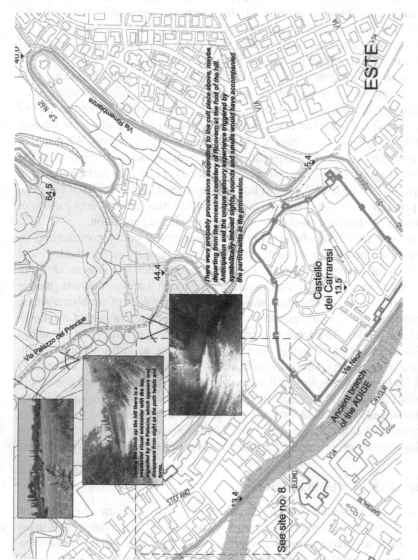

There were probably processions ascending to the cult place above, maybe departing from the ancestral cemetery of Ricovero at the foot of the hill. Anticipation and the unique sensory experience triggered by symbolically-imbued sights, sounds and smells would have accompanied the participants in the procession.

During the climb up the hill there is a sequential visual encounter with the top, signalled by the Palazzo, which appears and disappears from sight as the path twists and turns.

ESTE

Via Palazzo del Principe

Via Rimembranze

Castello dei Carraresi
13.5

Ancient branch of the ADIGE

See site no: 8

Figure 3.3 The community map of Este-Caldevigo and the Colle del Principe

Source: Copyright Sarah De Nardi

Apart from Archaeoclub members, several of my co-researchers had not personally encountered the material culture objects recovered from the site, and yet the place made sense to them as an archaeological and heritage landmark. How, if not through the imagination, hearsay and word of mouth? It is not just the material culture from the site that bridges the present and the past: an affective bond to place makes the site co-present within living communities and affords it a thriving place in their memory and imagination (McAtackney, 2015). Then again, some who had actively interacted with, looked after and stored the bronze figurines and other material culture from the site gave the objects private nicknames.

Therefore, I wondered which imaginations, perceptions and knowledge held by the community were part of an affective sense of belonging *in place* even more than *in time*. Many residents construe their perception and understandings of archaeological and historical features and sites their own way, which are often at odds with the specialist's normative and naturalising interpretations. What started as an exercise in landscape exploration with local residents culminated in the mapping of 'grassroots' geographical imaginations. Fieldwork participation and its maps (De Nardi, 2014a) merged into one another. These maps trace the emotional geographies of the everyday; their colourful visualisations form a mode of affective co-production as much as a tool for knowledge generation.

I start with a brief introduction to the site and project before commenting on the disconnect between community insights (especially when unorthodox) and the practice of heritage fieldwork – a tendency embedded in uneven and top-down institutional politics in Italy (section 3). Section 4 argues that imaginative and affective connections to place *make* cultural geographies of the Monte Altare landscape a form of local resistance to the academy's strictures and top-down normalising values. The local imaginary assigns diverse values and meanings to archaeological 'assets' that *come* from the site and, to institutional academicians, merely represent a quantitatively significant assemblage. Back in the early 2000s, landscape archaeology still seems somewhat detached from the contemporary social and cultural contexts, whereas cultural and social geographical studies were already concerned with the social value of heritage and the ethics of co-production (Cahill, Sultana and Pain, 2007). It comes as no surprise that our mapping exercise owes more to the latter set of subdisciplines than to the former.

Monte Altare: a site, many stories

What follows is an account of the 'life-story' of the Monte Altare. We may think of it as a collaborative, collective cosmology of the hilly site collated from many conversations and exploration with my local co-researchers. What is most interesting about the 'walking oral histories' we collected is that the local communities, when referring to the hill, show more than a respectful and caring attitude of stewardship towards their heritage landscape (see also Waterton, 2014). Residents almost refer to the hill as if it was a person, a beloved friend or relation. A degree of anthropomorphic investment in the hill (see a participant's following comment on the hill's likeness to a slumped, sleeping giant) sets it apart from your 'average' archaeological landscape.

My joint research with local avocational heritage experts identified successive chronological phases of occupation and human engagement with the Monte Altare landscape, drafting the development of place meanings through the millennia: the perspective of dwellers, conquerors, believers and antiquarians. We knew that at some point in the past the Monte Altare had been chosen from a cluster of local hills for 'special' attention. The Gruppo Archeologico del Cenedese (GAC) Archaeoclub's recovery of ritual figurines and hilltop-shaped bronze artefacts dating from the Iron Age (7th–2nd centuries BC) suggests a fertility cult concentrated on the upper slopes and top. Contemporary pottery fragments combined with a number of springs (now mostly dried out to a faint trickle in two places) might also indicate a water cult. Arnosti (1993) argues there was a shift towards an oracular cult in the Roman period: worshippers had their future foretold with numerical tablets. The GAC members believe a priest or priestess was in charge of divination, a long-lived and obviously popular practice that left behind the tablets which have been found in large clusters on the hilltop. Besides, locals with a knowledge of the divination tablets surmise this ancient pagan practice may be reflected in the contemporary memory/myth of the '*buss de la vecia*', 'the old woman's nook' in the local dialect. This is a rock crevice alleged to exist on the east slope of the Monte Altare that has not been definitely identified. Different residents locate this 'spot' in three different, hard-to-reach places, and no one has bothered to explore them, which has led to some logistical problems in drafting a heritage map of the area. The fact that the rock shelter or rock crevice exists suffices to place the old hag or old woman as a permanent, yet elusive, inhabitant of the hill.

After the decline of the Roman empire, the ambitious Christian Germanic warriors THE Lombards/Longobards conquered the area. At the foot of the Monte Altare, in the locality known as Salsa (*aqua salsa*, salty water, reminiscent of an earlier cult of hot springs) we find a sprawling Lombard cemetery which has yielded no end of treasure. A spectacular specimen of solid gold cross was recovered – which at some point ended up on display in the Late Antique gallery of the British Museum. Then, during the later Middle Ages, the Monte Altare became known as *Colum Maledictum* ('cursed hill', in Arnosti, 1993): the place name first occurs in a 1398 diocese boundary record. This nickname of the hill might reflect the chance recovery by locals of phallic bronze figurines – the idol-statuettes of their prehistoric ancestors. A distinct whiff of diabolical sulphur and heathen impurity likely led to the decision to construct two chapels in the 15th century in an uninhabited area of the hill. As nobody lived nearby, the only practical use of the chapels was that of colonising the abject space polluted by the so-called cursed artefacts.

The 'demonisation' and purging of the hill through the construction of the two chapels in the Middle Ages is in sharp contrast with a positive perception of the Monte Altare during the Renaissance. In the 16th century, the hill became the setting for a significant episode in the hagiography of St Tiziano, Bishop of Ceneda, and thus became part of a sacred geography blending faith, myth and legend in the imagination of local artists. Later, during the 19th century, the antiquarian diaries and letters of local scholar Francesco Trojer bear witness to a budding involvement by locals in their heritage: the antiquarian recounts lively interest in the excavation of a Roman cemetery, with residents from the working classes actively helping in the retrieval and storage of materials in the cathedral's crypt.

The hill, in the context of the province of Treviso, is significant in modern-day Italy: its peak bears a cross mourning the loss of lives in the First World War (see Figure 3.1). The powers that be picked the Monte Altare as the site to erect the Cross to the Fallen in the Great War. The area was the site of a decisive battle on the Italian front, and streets and squares throughout Italy are named after the town. In regional and national terms, the site's material culture is significant and well preserved. The story of the Monte Altare is a colourful tale told by many voices through many materialities. When embarking on my doctoral research in early 2003, I felt from the outset somewhat restricted by the initial focus on the prehistoric and early historic landscapes imposed on me by the strictures of the academic thesis. My community co-researchers briefed me about the landscape of the Monte Altare using a timescale that went from the prehistoric Iron Age (6th to 2nd centuries BC) to the Roman period (from the 2nd century BC to the 4th century AD) and beyond. Their framework also incorporated Late Antique (5th–8th centuries AD), medieval, post-medieval, modern (Second World War) and contemporary perceptions of landscape. After discussions with community archaeologists and interested co-researchers, a time-space holistic study seemed a better starting point for my investigations. I mused that, after all, heritage values cannot easily be untangled amid the web of overlapping meanings that make places. In our fieldwork entities such as place names, legends and memory, albeit intangible and possibly ephemeral, constitute as much a part of material culture as monuments and artefacts (Orange, 2010; McAtackney, 2016).

I chose this case study (see De Nardi, 2014a) because of its extraordinary complexity. The hill of Monte Altare has been experienced and understood in many different ways by the local inhabitants through time, from prehistory to the modern day. With the help of my astute co-researchers, I got to know this variegated landscape using a methodology which included landscape reconnaissance, ground photography, archival research, oral history, historical map research and place-name studies. Traditionally archaeological 'data' was complemented by the input of non-professional experts and residents (see also Orange and Peters, 2011 for an interesting Cornish parallel). The story I wanted to tell was shaped by landscape explorations with local people about a living landscape that accrued many memories, a network of meaningful places that had been created, discovered, loved, discarded, feared, rediscovered, hidden and reappreciated through time. The Monte Altare *had* also, I found, been subjected to "deliberate acts of forgetting and dislocation" (Chadwick and Gibson, 2013b: 8).

This complexity can and should be promoted by heritage managers on all scales (from the Sunday enthusiast to the professional) as a vital and fascinating aspect of cultural landscapes. Some of the concepts and concerns encountered in the review of current landscape practice that follows apply to the site of Monte Altare: here, we come up against issues of identity and self-representation. Alongside ideas of identity, we find a negation of the chosen self-representation by this heritage community which is perceived as marginal in the wider Veneto region. Our community heritage fieldwork – our walks, dialogues, discussions and interactions – resulted in lively, reflexive, multivocal, socially inclusive narratives and understandings which we mutually recognised, compared and accepted.

No place for dreams in heritage practice?

This section identifies some trends in recent and current landscape archaeology and heritage in Italy and beyond, which is, to an extent, still somewhat 'blinkered'. Simply put, ephemeral and intangible assets and facets of place are harder to understand and more complex to convey than visible ones. A cognate preoccupation with the reassuring "durability of material culture" (Jones, 2007: 3) detracts from the importance of ephemeral or intangible material culture: e.g. place names. Other aspects of landscape and place-making practices are overlooked in quantitative studies or regional surveys (Zerubavel, 2003; Orange, 2010).

The holy grail of inclusivity is not always put into practice in prehistoric landscape archaeology (see Tolia-Kelly, 2006; Watson and Waterton, 2015; Witcher, Tolia-Kelly and Hingley, 2010 for UK-specific case studies). Further, archaeologists usually work on a specific period in the landscape, attempting to filter out meanings and practices that distract from the interpretation of that era or phase of occupation or use (Tilley, 1994, 2004; Thomas, 1991; Kirk, 1992). Tourists visiting heritage sites might also tend to exclude modernity from their 'vintage' experience of place (Atia, 2010). The attempt to decant time-specific data and meanings has often obfuscated a wider holistic understanding of what place is and what place *does* (Solli, 1996). The Italian case study complicates matters further in that many scholars of landscapes and key stakeholders in, and managers of, heritage sites in that country still do not include contemporary perceptions, beliefs, opinions and the inhabitants' views in their interpretation of past and present cultural landscapes – creating the impression of studying 'palaeocultures' in a void (but see Sultana, 2007; Pickering and Keightley, 2012).

At the start of my doctorate I was dissatisfied with ways in which prehistoric and early historic cultural landscapes were portrayed in mainstream archaeological literature. Whereas historic archaeology and vernacular landscape studies are intimately connected with community identity (e.g. Alcock, 2002; Meskell, 2002; Riley and Harvey, 2005; Watson and Waterton, 2010; Iacono and Këlliçi, 2015), prehistoric archaeology still seemed somewhat detached from the social and cultural context in which we live today (e.g. Tilley, 1994; Jones, 2007). Those who theorised about human perception of landscapes within the postprocessual tradition tended to isolate the timeline they wished to investigate in spatial and chronological terms while ignoring features and landforms that did not belong to that period. For instance, some scholars of prehistoric landscapes seem to assume that Roman, medieval and modern individuals had become disembodied from their surroundings, that nature and landscape ceased being a form of consciousness and material culture for them (but see Alcock, 2002; Andriotis, 2008).

It is, I think, not just a matter of temporality but also of method. The Hadrian's Wall heritage landscape looms large in the local imagination due to its physical remains (Witcher, Tolia-Kelly and Hingley, 2010), but what if such traces do not exist? What other information might become 'data' if no structure remains? How reliable or even relevant is the archaeological imagination? Rather than dismissing informants' accounts as imaginative 'interpretations' – elaborate metaphorical accounts of a 'reality' that is already given – anthropologists might instead seize on these engagements as opportunities from which "novel theoretical understandings can *emerge*" (Henare et al., 2007: 1, my emphasis).

And yet stories about a place or landscape "move, emerge and affect in the very act of their telling" (Cameron, 2012: 588; Wylie, 2005). Similarly, oral history and traditions, as well as non-official maps, should not be underestimated as research tools (Riley and Harvey, 2005). Oral histories live in the "ambiguous space between here – the site of retelling – and there – the site of experience" (Cole, 2015: 4). When asked to reproduce the heritage landscape of Mount Caburn in East Sussex (England), in the context of an archaeological study, artist Carolyn Trant writes: "archaeology gives us at best only a brief and partial impression of the past, and I was determined to record what was still with us in the land, rather than trying to imagine a past by creating reconstructions of events with people in exotic costumes" (1987: 13). Trant's art imagined the past in a close relationship with the present: as *co-present*. She chose to engage with a dynamic non-absent past *in place*. Place is "central to our way of being in the world", and the task of the ethnographer (and archaeologist, I would add) is to "consider how she or he are emplaced, and her or his role in the constitution of that place" (Pink, 2009: 46). That is vital in any ethnographic venture but, as I hopefully show in the following, also for the archaeologist: an actor among many in the exploration of a site.

Archaeologists rarely include forms of material culture such as oral history, folk tales, historical maps, place names and myths in the story they tell – and this only where 'suitably reliable' backup sources are available. These ephemeral entities, many would argue, do not belong in the site catalogue or field report, but are still paramount to the interpretation of certain sites. Or rather, any given place is not likely be experienced in the same way by the same person twice, let alone by past persons and academics centuries or millennia after a mythical 'original context' of the site, place or landscape started to fade (Knapp and Ashmore, 1999; Waterton, 2005). For Rouverol (1999: 76) while we may disagree with a certain interpretation, "it is our obligation [. . .] to offer in our analyses conflicting interpretations, or what may seem to be paradoxical reflections or assessment".

A fascination with historic and archaeological landscapes is implicated with embodied modes of 'dwelling', belonging and rootedness as much as with scientific curiosity (Hamilton et al., 2006; Shanks, 2012; Ingold, 1993): this is important for the Monte Altare. Yet heritage professionals' perceptions and attitudes of the site, and the policy of management authorities based in Venice or Rome, could not be further apart from the reality of the local residents. Local residents involved in non-professional archaeological activities display an attachment to places as well as to the objects they find – for them, local heritage per se becomes part of the bigger picture (Castañeda and Matthews, 2008b: 2 ff.; Orange and Peters, 2011).

During fieldwork local experts guide and advise less knowledgeable members of the community. In so doing they have fun while at the same time showing a commitment to accuracy in the recording and knowledge transmission. As far as it is possible and realistic, local experts always make a point of spatially contextualising their finds in the field. They, the unofficial stewards, observe and record while also populating their tales in the field with personal episodes. The whole process of fieldwork becomes a social ritual, an adventure. Meanwhile the professional archaeologist is out there, a stranger looking in. So where do the Archaeoclubs and their local imaginaries figure in the bigger picture of heritage management? Locals understand places like the Monte Altare as mnemonic devices reinforcing a sense of local identity. By claiming the long history of place, remembering is an act of construction which makes the past meaningful in the present. Whether creating continuity (the Roman cult) or breaking with the past (the late medieval chapels), different acts of construction and destruction have shaped the Monte Altare's biography – that is, if we can ascribe this linear property to a landscape of such depth and complexity (see also Meskell, 2003). In the local area, non-professional archivists and historians discovered antiquarian diaries and letters in dusty attics and forgotten church archives, and they published these artefacts at their own expense in local journals. And the fact that they have been ignored, despite alerting the main universities of the existence of these documents, leads these part-time scholars to the impression that the cultural heritage of the area is of little or no interest to national- and regional-scale educational institutions.

> The Padua [University] archaeologists come and take away the bits of archaeology they are interested in. . . . While discarding everything else.[1]

Again,

> Funny to think that all these big bosses at the university and Superintendence in Venice know a lot about our prehistory and we have a Lombard piece at the British Museum in London, but our 18th and 19th centuries have become our dark ages! No one cares what happened then.

The latter comment was made by Samir, a Tunisian-born café owner after attending a lecture on the state of knowledge of local history and gaps in publication of historical documents. He and his wife were interested in local archaeology, occasionally participating in excursions on the Monte Altare. As Samir astutely noted, the significance of the Monte Altare's later phases is unpublished at a regional or national level and well-known and rehearsed among interested residents and local enthusiasts who did their own research on the topic.

Adding new knowledge of the past to their experience of 'home' makes the experience rewarding and productive. Local specialists and volunteers are keen to get involved, share knowledge and support specialists. However, any large extent of involvement is unlikely – often the overworked and, mostly, uninterested professional archaeologist disregards their input in the creation of knowledge, and as a consequence the locals are barely (if at all) quoted or acknowledged as sources of information in scholarly publications. This situation is not unique to the Italian northeast, nor is it a feature of this area alone. Up and down the country, committed members of Archaeoclubs actively engage in their heritage by devoting their spare time to field walking, analysis, washing and storage of unwanted or surplus crates of material and organising events with schools and local conferences open for all to attend. I have myself presented at such open events, and in my experience the atmosphere and participation are always spontaneous and dynamic, and free admission means everyone can join in the discussion without fear of being judged on their knowledge or on the basis of their language skills – dialect is widely spoken in the region and generally preferred to Italian.

After several interviews and focus groups with members of local Archaeoclubs it became apparent that the origins of many historical societies and Archaeoclubs in modern and contemporary Italy have as much to do with attachment to place as with scientific curiosity per se, with the latter being the main drive for academic and administrative archaeologists based in Venice and Rome. To busy professionals, places like the archaeological landscape of the Monte Altare constitute just an inventory number in a marginal regional site database.

In any event, this enduring dichotomy determines the contrast between scholarly archaeology in Italy – within the realm of the universities and museums (in the urban contexts of Venice and Padua) and of national-scale institutions mainly based in Rome and primarily concerned with artefacts and sites estranged from their local cultural landscape settings. Practitioners and academics need to ask the right questions, to the right people, and incorporate that feedback into their work in order to empower communities (McDavid, 2002; Grey, 2008; Vuyk, 2010). Without empirical evidence to measure the public's support for and interest in cultural heritage, the contexts in which fieldwork is practiced are vulnerable to public policy changes and the broader impacts of economic austerity, be these community projects or within museums (see also Gallaher, 2016). Waterton (2005) among others emphasises the need for, and ethical responsibility of, archaeologists involved in the presentation of their work in the public realm to understand, respect and value the interpretations of the past by non-professionals, without imposing their agendas on the communities they work with (McDavid, 2010; Horning, 2013). An inclusive approach "necessarily involves multi-component, polyphonic perspectives, requiring an active effort to seek out and promote voices that are harder to hear" (Mickel and Knodell, 2015: 253; see also Meskell, 2002).

Local heritage associations and communities of non-experts consider sites, historic landscapes and objects part of their past, present and future identity. The fact that many of the artefacts and sites had been discovered at a local level by non-professional enthusiasts out on a Sunday walk in the hills makes the bond of stewardship between locality and heritage even stronger: these people associate the object intimately with the location in which they found it, the circumstances in which it was come across, and they live and relive the event in an embodied, sensual manner that is often lost in the post-retrieval catalogue and analysis work (Figure 3.4).

Heritage communities: making places and enacting history(ies) 'in the field'

I believe that imaginative 'sense of place' plays a paramount role in local perceptions of heritage landscapes. Despite recent criticism (e.g. Wiley, 2007; Johnson, 2006), elements of a phenomenological inquisitiveness and the imagination inform the birth of and involvement of local communities with archaeological and historical heritage (Shanks, 2012; see also Gregory, 1994; Orange and Laviolette, 2010; Orange, 2010). What follows is an account of our site explorations, privileging the imagination and the senses over 'hard' scientific data. The two, I argue, can coexist peacefully and create holistic understanding of what heritage *feels* like – what it feels like to have 'ancient remains' near you.

So, this is what we did. I invited local residents and their friends and families to participate in fieldwork at the Monte Altare and a number of other locations. A team of ten including local archaeologists, historians, place-name scholars and interested residents were happy to take part in this pilot phase of the project (see De Nardi, 2014a). We met at a cafe bar in the centre of the town, where by mutual agreement we verbally negotiated a route of navigation towards the top of the hill and set off. The pilot group was formed of ten people, including myself. My co-researchers at the Monte Altare needed no maps to navigate the site as they were all thoroughly familiar with the landscape.

Going up on the Monte Altare is a bit like visiting an old friend.[2]

The team members ascended the steep, verdant slopes of the Monte Altare and explored the topography they knew so well. I recorded our conversations, and every so often I asked co-researchers for their understanding and knowledge of particular rocks, passages, vistas and familiar sounds and requested them to take mental notes of their experience that day. At the point of departure, the fieldwork participants discussed the landmark of the Monte Altare. They agreed that it looks striking, as it varies in appearance depending on where you are standing and looking at it (Figures 3.4–3.5). One member of the team, librarian Carlo, reported (Figure 3.5):

The way you see the hill, it . . . varies so much. If you stand and look at it from the southwest, on misty days, it almost looks like a sleeping giant, slumped on a bed of rocks. It's fascinating.

Insights like Carlo's, one would argue, are of little use to the professional archaeologist concerned with the factual recording of data – and yet to him, how the hill looks is important.

As it turns out, the route up to the hill presented a sequential opening up of vistas and landscape features and narratives: there seemed to be a certain order in nature, or rather, we got the distinct impression that, on the hill, nature constituted a structure in its own right, guiding our steps through the familiar landscape.

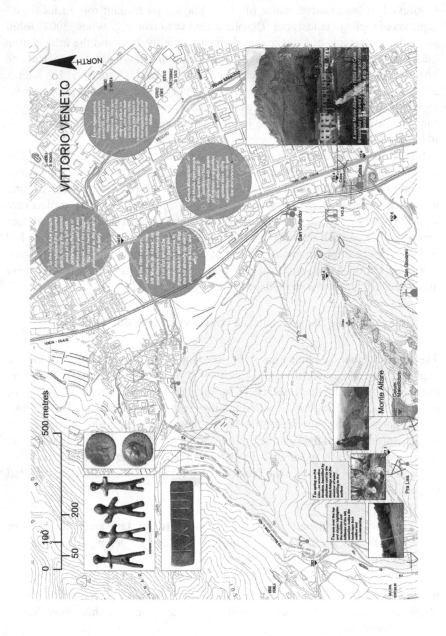

VITTORIO VENETO

NORTH

500 metres

0 50 100 200

River Meschio

San Gottardo

San Giovanni

Monte Altare

Pra Loto

A Roman Mortar Altar *Brazzolo*, carved from a human and horse *torso, with a human at its foot*

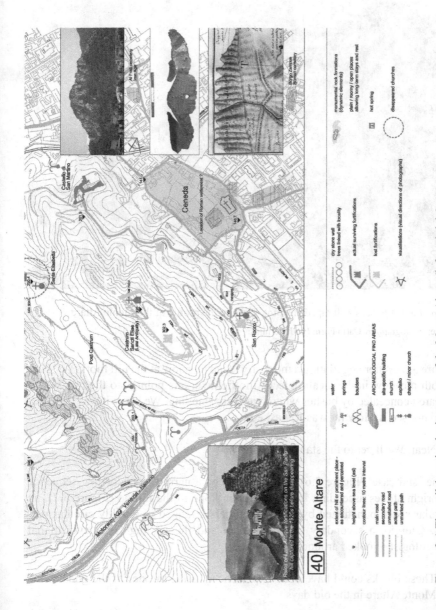

Figure 3.4 The collaborative heritage map of the Monte Altare (left and right)

Source: Copyright Sarah De Nardi

Figure 3.5 Carlo sees a slumped sleeping giant in the profile of the Monte Altare
Source: Photograph by Carlo Forin (with permission)

(Figure 3.6). After negotiating a modestly steep meadow at the hill foot (*Pra'Liss*, smooth meadow in the local dialect), the fieldworkers pointed to the top, which appeared concealed by bushes where we were standing. We knew it was there but could not see it. How far was it, I asked?

Near. We'll get to the staircase soon.

The 'staircase' (Figure 3.6) turned out to be a narrow corridor-like sequence of upright stones ascending from the upper slope to the hilltop (Figure 3.7), an irregular space punctuated by square boulders arranged in roughly circular patterns. Giorgio, a secondary school teacher and the leading GAC avocational archaeologist – looked around and grinned:

These blocks could have been experienced as the thrones of the gods of the Monte Altare in the old days.

A co-researcher in the second team of explorers took me aside and confessed that the top of the Monte Altare was the spot where she and her then fiancé held their romantic rendezvous:

Figure 3.6 The natural staircase leading to the top of the Monte Altare

Source: Photograph by Eva Favero (with permission)

He used to live just at the foot of the hill. I have not been able to look at the hill the same way again, you know?

The local avocational archaeologists and interested enthusiasts recounted the personal experiences of their discovery of the artefacts, and Giorgio admitted to attaching affectionate nicknames to some of the objects they had found.

Figure 3.7 The thrones of the gods on top of the Monte Altare
Source: Photograph by Sarah De Nardi

> I call them Smurfs, because they are roughly the same size as the Smurf figurines, you know.

Touch can be a medium for "annihilating time and space and establishing an imaginative intimacy with the former possessors of articles" (Classen and Howes, 2006: 202). Never was this intimation more accurate than in this case. Thanks to Giorgio's insight I was able to immediately picture the size of the artefacts without so much as looking at them. Giorgio and Carlo tell the story of when they found the Iron Age Smurfs during fieldwork. One was the middle of the rocks up top and in cracks in the dry soil at the foot of the hill.

> Dull, oxidised, but what a sight!

More co-researchers volunteered personal stories, provided fieldwork anecdotes and exchanged memories and impressions. Mirto and Sandra offered tales of local legends pertaining to the hill and its environs; this well-read couple conjured up a vivid mixture of medieval hagiography and Catholic superstitions.

GIOVANNI: They built not one, but two chapels up there near the top when they found the phallic statuettes [. . .] to exorcise the demons.

MARIA: What happened to the chapels?
GIORGIO: Ruined.
EVA: But when?

No one could answer her. It seems that, once they no longer served the purpose of guarding the hill from evil, people dismantled the buildings for masonry. It is unclear when the dismantling took place – but the remnant of the churches lingers in the historical memory.

Pensioner Alberto told the local stories of the First World War as it was fought on the hills by his grandfather, and Giovanni volunteered colourful anecdotes of local 19th-century antiquarianism. They take pride in their work and are eager to learn more, to engage in a constructive, mutually beneficial dialogue with the academic (see the following). The collaborative element of fieldwork proved most rewarding, so much so that local archaeologists, residents and I (the outsider, the scholar) jointly contributed to the creation of experiential 'heritage' maps of several locations, in which we inscribed on the landscape our emotions and impressions as well as the location of archaeological finds and structures.

A pattern consistently emerging from fieldwork and interviews in northeast Italy is the awareness that locals engage with their heritage as a matter of pride and as a symbol of belonging. Here, heritage is experienced through the senses – as the participants identified places of artefact recovery, the location of trenches, legendary spots and curious landscape elements, they also inspected the local topography for traces of illegal excavations and disturbance. Everyone admitted they thoroughly cherished their role in maintaining the Monte Altare hillscape clean, tidy and accessible. But what does the Monte Altare represent for members of the local community who are not familiar with its history and heritage association today? A chat with a resident I met in a cafe after the first fieldwork session turned into a spontaneous interview: after admitting that he knows little about the archaeology of the site he said:

> I work outside Vittorio Veneto . . . . I commute daily by car to Padua. When I am homeward-bound, on the motorway, the Monte Altare is the first landmark that tells me I've made it home, safe and sound.[3]

He was curious about our project and joined our second group of volunteers the following month. Other people who were in the cafe agreed, with one volunteering,

> You know where you are as soon as you see those rugged peaks.

To many, then, the Monte Altare is largely a landmark of 'coming home', the first recognisable landmark from the motorway. Yet in the archaeological literature this special and visual attachment to the hill's shapes and contours is nowhere to be found; the heritage landscape of the Monte Altare is the archaeologist's playground, and the meanings attached to the place are remote and academic and belong to the specialist. Going by the scarce academic publications on the site, anyone not familiar with the local cultural landscape would assume that the Monte Altare landscape is a relic, a fossil representation of the ancestral past. They would not suspect that the hill is a lived in, vibrant crucible of meanings, memories and contemporary associations. In summation, fieldwork at the Monte Altare site comes to life across a multivocal and brecciated landscape in which local perceptions past and present, situated knowledge and local myths are a major part of fieldwork and research and mnemonic aids to learning about one's past and heritage (Jarrett, 2013).

Visualising what? Implications and yet more questions

The vibrant multi-period cultural landscape of the Monte Altare plainly shows that there are multiple ways to appreciate a heritage landscape. Not all ways are shaped or dictated by the presence or visibility of factual historical and archaeological data (e.g. Gambacurta and Gorini, 2005). We now know that experiencing the past is as diverse as experiencing place (see also Stroulia, 2016; Witcher, Tolia-Kelly and Hingley, 2010). As a researcher in the field, I discovered that an exclusive interest in only one aspect or historic period of a landscape may ignore and occlude other histories it contains. In their role of residents and co-researchers, participants' interpretation and awareness of the past inscribed on the hill and its environs possess a timeless chronology and un-Euclidean geography which are accretional, fluid and imaginative.

The knowledge and understanding of places like the Monte Altare draw on affective stances and familiarity with the natural setting of the heritage site, as well as a local fascination with and knowledge of the stories and traditions surrounding it. A plurality of views and opinions about this place characterise the research findings, alongside some shared themes that emerge as significantly from co-research and community field walking. Drawing on their situated knowledge and community-led views and perceptions, the residents construct and experience their own heritage landscapes, in their own terms. Tapping into the complex and sometimes unexpected local imaginaries is a viable way to interpret and visualise the complex narratives of landscape and is necessary to develop a long-term sense of place, where locales, impressions, people and things build on one another and grow out of natural and architectural spaces (Gregory, 1994).

My main argument here is that we need ways to incorporate imaginative geographies that do not conform to the format and content promoted by official heritage bodies. An acknowledgement of the affectual histories permeating a place and affording agency above and beyond the 'taming' of archaeological science requires that we stop compartmentalising eras of occupation; we might then also focus on the emotional longue durée unoccluded by specific historical or archaeological periods.

Elsewhere (2014a) I have proposed a mapping strategy to draw together subjective and objective facts and notions about sites; maps can become a means to empower community and disrupt top-down values (Del Casino and Hanna, 2006) through their fluid becoming from inception to production, use and duplication (see also Gibson, 2001). Through joint fieldwork and participatory mapping practices we might foreground interpretations and understandings of heritage from the perspective of individuals and communities intimately acquainted with their heritage landscapes. The maps and stories thus produced would then represent a viable alternative to the hegemonic meanings imposed on communities by 'outsiders' (bureaucrats, experts) treating any given area as the object of academic study without establishing a dialogue with the locals and remaining ignorant of their perceptions and knowledges about place. I would argue that this lack of dialogue has led the local community to identify a discontent and a desire to be heard (as voiced by Samir), to have a say in the management, presentation and interpretation of their past, present and future heritage(s).

The fact that certain landscapes are long-lived – and long-loved – indicates that places do matter to them, that places have a personality and an agency and that sites like the Monte Altare are an essential component of group identity and pride. The main finding of this study is that Archaeoclubs members and volunteers appreciate and explore their heritage from within, building on their sense of familiarity with a well-known landscape in their understandings of local histories, lore and traditions – familiar surroundings make sense of old and modern tales and define the inhabitants of the localities around the Monte Altare (Costa, Fadalto, Ceneda) as 'belonging' there. In this area, 'heritage' is largely perceived as a social group's joint 'dowry' and as a vibrant and inclusive reminder of past lives, rather than yet another tourism or economic resource to be negotiated and managed by professionals.

During fieldwork we also noticed that the so-called disciplinary divide among archaeology, history, ethnology and geography is untenable at a local, non-professional level – and that local heritage volunteers do not spontaneously 'classify' their expertise and knowledge of meaningful historic landscapes into neat academic departmental labels or chronological periods. At the other end of the spectrum academic research agendas approach and analyse heritage landscapes such as the Monte Altare 'from without' – the resulting insights and narratives spanning from professional fieldwork lack inside knowledge and are often unaware of locally constructed and negotiated understandings of place.

This chapter has sought to foreground the fundamental, no, essential role of local communities in the construction of an all-round 'sense of the past' (see also Atalay, 2012). Local knowledge and the 'emic' meanings attributed to heritage landscapes by those who inhabit them, however unorthodox or imaginative, are vital in shaping and nurturing the awareness and acknowledgement of local landmarks and identities. Material culture at such complex sites simply has to include the aspirations, desires, emotions and memories of many. This imperative resonates with ethical practice in knowledge production of cultural and heritage landscapes through research and fieldwork across the disciplines of archaeology, human geography, heritage studies and local history (Jones and Garde-Jansen, 2012). It would be desirable if, in the not too distant future, Italian heritage bodies implemented and followed policies of social inclusion already well established elsewhere in Europe; a good example might be the Heritage Lottery Fund in the UK's mission to "listen carefully to the changing ways in which an evolving society values the past".[4]

Summary points

- A multi-period map of the area brought together multiple viewpoints concentrated around memory and perception of the site
- The map was the outcome of dynamic site visits and as such encompasses sequential, summative and point-of-view-specific referents relating to movement on the slopes of the hill
- Natural and cultural elements form part of a single experiential sensory realm
- Present and absent elements of the landscapes blended together
- Memory and material culture found space on the map side by side
- Some co-researchers had not personally encountered the material culture objects in fieldwork, but the place made sense to them as an archaeological and heritage landmark

Postscript

This chapter is dedicated to the memory of Professor Steve Watson – a champion of community heritage and grassroots understandings of heritage places.

Notes

1 Mirto M., interview, 25 August 2004.
2 Maria V., interview, 30 August 2002.
3 Angelo B., interview, 23 July 2003.
4 (HLF 2006).

4 A town, divided

Mapping the tense imaginaries of the 1943–1945 Italian civil war

Mnemonic communities: a reflection on (infra)humanity

"Why are Italians still so obsessed with the civil war and the Resistance? Why do they keep arguing about it?" These questions were formulated by a student in 2015 – questions that are entirely legitimate, that are clever and to which there is no answer. No answer I could think of, at any rate. The scale of the phenomenon, unarguably, has had an impact on its endurance in popular memory. Its pervasive haunting has never really paled, and reconciliation is still far from reality. Taken together, these matters of scale and haunting endurance are central to this book's story and as such deserve unpacking.

Italy entered the Second World War alongside Hitler's Germany in 1940, strong in the ideological alliance of Benito Mussolini and the Führer. Following a disastrous war campaign in which the German partner proved to be a master rather than an ally, Mussolini was ousted and exiled, an armistice with the Allies was reached and Italy split from Germany and started fighting against its former ally, turning from friend to enemy of the Reich overnight. After Hitler sprung Mussolini from prison and installed him as the head of the Repubblica Sociale Italiana (RSI), a puppet republic in northern Italy, the country was split in two. The south was in Allied hands, with the centre and north at the mercy of the Fascists and their Nazi masters. Between 1943–1945 the ruthless German occupation of Italy triggered a hitherto political – if often bloody – conflict between Fascists and anti-Fascists into full-on civil war.

In northeast Italy, the first partisan bands were formed after Germans started pouring into towns and cities – Vittorio Veneto, a town in the province of Treviso, for instance, was occupied on 14 September 1943, a mere week after the Armistice (Della Libera, 1988: 18). Towns and villages throughout the region became unsafe for anyone opposing the authority of the Reich and their Fascist aides, or for anyone not responding to recruitment roundups in the new armed forces of the regime, so that rebel fighters had no choice but to 'take to the mountains', to escape to out-of-the-way places, safer in their remoteness than in their hometowns (Della Libera, 1988: 20; De Nardi, 2016). By the end of 1943, there were around 9,000 partisans in the Italian north (Ginsborg, 1990: 16).

In terms of scale, the anti-Fascist Resistance movement was significant in military and political terms: some have claimed that 300,000 Italians were involved in direct action against the Nazi-Fascist Salò regime by April 1945, with many more supporting partisans with provisions, intelligence and shelter (see Bocca, 2008 among others). The Allied authorities at the time, and most historians to date, agree that the Resistance made a critical contribution to the eventual retreat of German forces by early May 1945 (Cooke, 2011). Without partisan action, Italy would have been much harder to deliver from the double stronghold of Mussolini and Hitler's armies, and a priceless strategic partnership in terms of logistics and intelligence would have been lost to the Allies.

As for its haunting pervasiveness in Italian consciousness, the phenomenon was as momentous: as German troops (with enlisted Italians) and Mussolini's Fascist forces battled with the partisans, Italians often found themselves fighting each other through conviction or coercion. In that war without a front, agency and identity enacted ephemeral boundaries between persons and groups. Civilians were subject to random violence, be it Allied bombing or several infamous massacres by retreating Nazis (Klinkhammer, 1996). No one could be fully trusted; Resistance activists and Fascists and Germans perceived the other, each other, as dangerous unknown quantities. Many among the former Resistance fighters and activists – especially those individuals of strong Communist leanings – resolutely negate the occurrence of a civil war. In their eyes, to admit to a 'civil war' looks bad for Italy, long perceived as the hybrid half ally/half foe of the Anglo-Americans.

At the grassroots level, which is the context for much of my research on the Italian war, I have instead found that most people prefer not to talk about the infighting between Italians because the memory of the war events still hurts. A civil war is a civil war, they say. When asked to elaborate on the meaning of that statement, an activist called Roberto replied that it is one thing to hunt down and kill Germans, foreigners who turned on you and were burning down your homes and your villages, raping and killing your family members, all the while 'barking orders' in a language you did not understand. He assured me that the invading German militias did this and much worse, although, as I argue elsewhere (De Nardi, 2016), several also behaved humanely and were kind to civilians. Sometimes, Wehrmacht soldiers even directly joined the Resistance fight against their own. For Roberto it was another thing to slaughter each other in a fratricidal war where everyone was against everyone.

Processes of infrahumanisation may be a key to decoding the maelstrom of difference and the minefield of blatant discrimination animating the popular imagination and remembrance of individuals and families who identify(ed) with either Fascists or anti-Fascists during the Second World War. Infrahumanisation theory posits that unless we perceive others to be fully human, capable of experiencing the full spectrum of human emotions, they are not like us (Leyens et al., 2000). We can harm them and deride them, and we will allow ourselves to believe they are not worthy of decency or respect. The affordances of such a stance may be unspeakable in a conflict situation (Curti, 2008). To this end in Italy, an avoidance of constructive confrontation between two factions (the Fascists and the anti-Fascists) who believe the other to be infrahuman leads to an unwillingness to acknowledge any degree of culpability of one's own 'side' in acts of violence. The activists and civilians during the years 1943–1945 were disjointed in many 'ingroups' and excluded 'outgroups', (Others) hopes and fears, then, may have been discredited as less important than one's own (see Haslam et al., 2005). I have grappled with this contentious issue in my first monograph (2016) and do not intend to return to the argument at length here. However, the positioning of the other and the other's memory is crucial to a mutual understanding of the past, of a shared experience and of the heritages that are created on both sides of the political and cultural divide in Europe.

Italy is not, of course, alone in the realm of enduring divisive memory. James V. Wertsch (2008) has explored what he terms blank spots of Communist memory in Russia, focusing on the secret protocols of the Molotov-Ribbentrop Pact of 1939 as a case study. Taner Aksam (2007) interrogates Turkish responsibility in the Armenian genocide and its hidden depths of wilful forgetfulness: different perspectives embroiled in a difficult heritage with many versions and declinations. Audrey Horning (2013) and Horning and Breen (2017) and Laura McAtackney (2014, 2015, 2018) have raised crucial points on the interactions of professional (archaeologists etc.) and community members in socially divided areas such as Northern Ireland (see also Cashman, 2006). The adopted stance in this research is usually to engage, sensitively and inclusively, in order to promote more nuanced versions of the past that unite rather than deepen the chasm. Material traces intermingle with memory to make sense of the world in the past and in the pervasive present (Till, 2005; McAtackney, 2016). Injured buildings echo the injured bodies of people caught in the mangle of conflict.

But what if we have no traces to build on to create cohesion? Due to the very nature of the organisation and strategy of the Resistance, material culture is scarce. The European freedom fighters had to leave no traces of their passage lest the local populace be punished or suspected of collusion with the 'rebels'. The partisans had to learn to disguise their presence wherever they went. What do we make of often-deleted traces today? This echoes McCarthy's anguished question, "How do we place those negative aspects of our culture which we indirectly and persistently sustain?" (2017: 61). In some sense, the following experiments in time-place exploration served to exorcise the pain and shame of fratricidal violence without erasing it from the social-cultural landscape. The decision in the course of this project to keep veterans from the two opposing political factions separate in fieldwork was not due to hesitation in front of controversy, but out of respect for the ageing individuals in question . . . and therein lies the main and most blatant limitation of the Italian-specific experiments that follow. The attempt to navigate and make sense of 'place' still divides and separates witnesses from each other. In my respect for two elderly individuals, I have perhaps fostered separation and division more than most. The cathartic exercise of mapping with the wider community is but an act of atonement for my shortcomings.

This failure to engage is not wholly my fault. A reluctance to participate, to really establish a dialogue in honesty and mutual respect between Fascists and anti-Fascists, is reflected in the literature. When post-war European ethnographies exist, they mostly focus on the traumatic, backward-looking qualities of warscapes and post-conflict geographies through the lens of mourning; the tangible (i.e. architectural and monumental) heritage of conflict experiences has been problematised as contested, dissonant and unwanted heritage (Ashworth, 1996; Rico, 2008; Larkin, 2012). It is still the case that studies of global conflict perception, while insightful, tend to focus on the political and the discursive and somewhat neglect the embodied, gendered and affectual sides of war experience (Petersen, 2005; but see Curti, 2008, Mookherjee, 2015, McAtackney, 2018). Other ethnographies of war (see Cappelletto, 2003; van Beschoeten, 2005; Focardi, 2013) exclude the materiality of the everyday and the material catalysts of memory from their accounts. Instead, in my Italian project my co-researchers and I have used oral sources to complement archival data and to gauge the extent of involvement and participation in the Resistance in northern Italy. During interviews, most veterans physically relived the relentless anxiety of the period; crucially, their memory is often linked to bodily sensation and the grounding of the body in physical space: the loneliness of a road at dusk, the constant hunger, the discomfort of hiding in flea-infested holes in the ground, the misery of night watches (De Nardi, 2015).

> On the Greek front you always knew where the enemy was; over here you never knew where the enemy was, where he was hidden – you were never sure.[1]

But who are the mnemonic communities who 'do' the remembering? How do they position themselves, and how do their memories interact spatially and affectually, becoming 'reluctant heritage'?

The Italian context

The Italian Resistance movement remains a highly controversial episode in that country's history (Portelli, 2003; Pezzino, 2005). Shades of in-betweenness in political and moral choices existed everywhere in Europe, but in Italy these ambiguities were even more prominent (Di Scala, 1999). The 'popular myth of the Resistance' (see Battaglia, 1953) developing after the war served as a mechanism to overwrite violence against the Fascists, accepted as necessary to the goal of Italy's liberation. Moreover, several individuals on both political sides (Fascist and anti-Fascist) who perpetrated war crimes escaped justice following a Presidential Decree of 22 June 1946, known as the Togliatti amnesty (Ginsborg, 1990: 102–104). What is interesting to note here is that, following decades of dominant pro-Resistance remembrance politics (25 April, Liberation Day, is a national public holiday), the anti-memory of those who did not participate in or were against the Resistance has been occluded (Mammone, 2006). In the bitter aftermath of the war, the losing Fascist side was mourning its dead in private, through do-it-yourself memorial sites. Only in the 1980s and later were the Fascists bold enough to resume interventions in public life – for instance, gathering on the anniversary of significant events and, often, complaining to town councils about dedications of public spaces to the Resistance.

In Italy, as in wider Europe, there is still reluctance to engage with the violence engendered by the conflict, yet the lives and deaths belonging to the Second World War feel close, haunting families and social groups; trauma is still fresh in many displaced or massacred communities in central and northern Italy, although not always in the public sphere (Antze and Lambek, 1996). After all, "there are always perpetrators and sufferers and their perceptions inevitably differ radically" (Logan and Reeves, 2009: 3). But the embodied nature of these wartime events means that they are not long-forgotten abstractions: they are present. In what follows I use the term 'affect' as a 'process of embodied meaning-making' (Wetherell, 2012: 4; see also Ahmed, 2004) that is profoundly social and central to the experience and to the negotiation of wartime memory and the collective historical imagination.

In the town of Vittorio Veneto in the Veneto region of northeast Italy (as in the rest of the country) controversy over the morality of the civil war and Resistance movement abounds (see mainly Pavone, 1991 [2015] but also Ballone, 2007). My own family bears witness to the ambiguity of what being an Italian means.[2] Doubts and subjective conjectures about the culpability and responsibility of the partisans smother this difficult conversation like a gag. The descendants and families of Fascists still live in the town, and they share the same spaces with the descendants of the anti-Fascists and the armed partisans who perpetrated violence towards fellow Italians. It has become almost farcical: every time the town administration (*Comune*) dedicates a park bench or a square to the Resistance, someone will almost inevitably come forward protesting that so-and-so in this or that brigade murdered their Fascist father/grandfather/great uncle.

Going back to my student's question that opened the chapter, I now believe many Italians are haunted and fascinated by the idea of a civil war in their towns and villages because it left so few tangible traces. There are no official monuments that explicitly commemorate the killing of Italians by other Italians in the fight of 1943–1945 – but then, monuments to the victims of pre–Second World War Fascist violence are also not visible. There are several official monuments commemorating the successes or mourning the loss of lives in the Resistance – but it feels different. Seeing official displays of public memory celebrating the winning side is not the same thing as reviewing the private chagrin of those who lost loved ones to a tragic fight between ideologies.

While a morbid and obsessive narrative on both sides pervades the historical and, indeed, the geographic imaginations of many Italians in the north and centre, these feelings and fixations are not connected to specific sites or to visible and tangible traces. Secrecy, marginality and, in some cases, the forbidden nature of some of these stories may have elevated them to the status of private mythologies. Amid an ontological divergence of 'display' and conceptualisation, the official remembrance of the war in Italy is epic, sterile and formulaic. It cites numbers, fighting bands and battalions, but does not convey the horror of blood and tears of conflict. "Academic scholars and armchair strategists should try to figure out step by step *how action takes place*" (Salvadori quoted in Tudor, 2004: 21, my emphasis). Walking memory explorations of towns in northern Italy may serve a different purpose: they seek to unlock a different side to the story. To bring to light different experiences and to expose the wounds in order for them to begin to heal. They may unbury silences by engendering conversations.

The questions at the forefront of my mind as I conceptualised the maps with members of Istituto per la Storia della Resistenza del Vittoriese (ISREV), the local Resistance studies institute, were the following: how have ambushes, killings, reprisals and denunciations been absorbed in the vernacular and remembered today? Sabotages, raids, desertions and executions did not take place as the impersonal documents or war diary entries we encounter in the archives today. Paramount to my interpretations and even design of the research was local historian Pier Paolo Brescacin's ambitious (and brave) two-part historical and anthropological essay on the violence perpetrated on both sides during the difficult years from 1943–1945 (Brescacin, 2012, 2014). Brescacin had skilfully demonstrated that these were real-life distressing fragments of lived experience: as all experiences of conflict, acts of violence happened in the battlefield of city streets, hillsides, remote villages and treacherous woodland (Cappelletto, 2003; Navaro-Yashin, 2009; De Nardi, 2016, 2019). Most importantly, these places overlapped (and still do today) with an everyday geography of home. I started from the fundamental concepts of *embodiment and emplacement* as a basis for the initial fieldwork design.

Remembrance is a dialogue

Memory is more than inscribed on place and bodies: in its far-reaching affectual potency, memory is embodied, haunting and enacted in the everyday acts of inhabiting and navigating cities and towns. Memory is presenced by people's and non-human agents' very existence and movement in those spaces (Drozdzewski, 2015). Sometimes the spaces under the geographer's lens are ghost towns (i.e. DeLyser, 1999; Navaro-Yashin, 2012) where absences are stronger than presences in the imagination of the visitor. Oftentimes the spaces where memory lives compete with contemporary preoccupations in busy, vibrant urbanscapes (Belcher et al., 2008) Whether exploring presences or ghostly absences (Meyers and Woodthorpe, 2008), geographical ethnographies are concerned with understanding the forces that make and populate places (e.g. Wertsch, 2008). Tangible and intangible materialities need to be made to speak to each other, but they are always already morally compromised and compromising.

Sites of wartime action colour the local imagination in many European cities and towns. The visceral quality of in situ memories brings them to the fore in daily experience in a disquieting spectacle of contemporary, present affective resonance. *How action takes place.* As I flipped through Tudor's slender compendium of British Special Operations Executive (SOE)[3]-related memories of the Italian Resistance (Tudor 2004), this phrase stuck insistently in my mind. How might we think of the way that action *makes* place too? What if we chose to engage the materiality of historical events? Battles, events and people *become* or generate their own material culture: these elements are part of the same existential and experiential entanglements. Above all, however personal, encounters with remembered or imagined violence do not exist in a vacuum; like memories, wounded landscapes are populated with "the lives and biographies of those who have lived before us" (Chadwick and Gibson, 2013b: 13). Memories are context- and audience-dependent and are "prompted by the availability of others" (Kavanagh, 2000: 15). On the other hand, imagination can be a lonely phenomenon taking one by surprise, but it may also trigger powerful feelings and associations.

Understanding and weaving together fantasies, fabrications and recollections may create a holistic conversation between parties, affording a sharing of conflicting meanings and experiences. These may then be better grasped and ultimately visualised through deep mapping (Thomas and Ross, 2013; De Nardi, 2014b). An open-ended, non-authoritative act and process might bring to life some of the less conventional stories in places affected by war and conflict (see Nordstrom, 1997). The concept of stories is inextricable from the agency of energies and forces channelled by, and conceptualised through, affect (Ahmed, 2010). Being in place, or revisiting places, alone or with others, triggers memories of disruptions to everyday dwelling practices and violent events at the location where something happened. When we enter a place where something memorable occurred, we tend to shift from a private, intimate memory to a shared, situated, physical memory. This is how war memory mapping *works*. By blending historical data, photographs, memories and stories, these maps can open up hitherto hidden or marginalised 'townscapes' of war and betrayal unfolding in the city streets.

In this chapter, I specifically reflect on how the interplay between the 'hidden' anti-memory of the civil war – the memory of violence perpetrated by the winning side, the Resistance – and the imagination of wartime events in the town of Vittorio Veneto creates troubled and troubling overlapping geographies of home and belonging or exclusion. Fieldwork, walking and mapping attempted to disentangle wartime atrocities from the urban fabric where they have long been silenced or pushed to the margins because they attract attention to partisan violence. Although the project actively encouraged both partisan veterans and Fascist veterans and sympathisers to enact their geographies of wartime events freely and without judgment or prejudice, the sites of forgotten violence have prevailed. It is as if, when asked to map out their emotional geographies of the conflict, both sides responded by ignoring the monuments to the Resistance and taking me, instead, to the small-scale, concealed and marginalised loci of their suffering and fear. What interests me here is the interplay of the imaginative and the remembered in the creation and rehearsal of such spatial stories.

To repeat the leading questions that would shape the mapping experiments: how have acts of ambushes, killings, reprisals and denunciations been absorbed in the vernacular and remembered today? How are violent and subversive acts reinvented and reimagined in the vernacular? Thus, my participants' impressions draft a spatial mythology of danger and violence which reflects real and 'doctored' events and facts (De Nardi, 2014a, 2016). In that town, traumatic and mundane meanings alike emerged and clustered, lingering like ghosts in city streets and at crossroads in the aftermath of the Second World War. I first had to comprehend what kind of triggers 'set people off' as they encounter places of violence: are emotions and affects triggered by memory, the imagination or both?

I then go on to discuss how these ephemeral yet pervasive affects (postmemories or dreams?) might be mapped out. Mapping can be a kind of archaeological investigation of buried, deep-seated meanings and affects otherwise imperceptible to us. Emplaced and stirring, the memory of ambushes, executions and violence perpetrated during the war in Vittorio Veneto is inscribed in street signs, monuments, atmospheres and tangible and intangible materialities that encompass public and private acts of commemoration. Fractured memories and (often) inflated or made-up revenge tales compete for our attention: where the losers (ex-Fascists) lay claim to sites of martyrdom, the winning side (the Resistance veterans) shows off their accolades. Finally, I articulate and situate local wartime recollections and (possibly) fabrications through three interwoven story-memories present on the map. These memory-stories were enacted or projected by local communities onto different areas in the cityscape of Vittorio Veneto. I lay out the three memories in chronological order to orientate the reader.

It is now time to turn to the traces, whether remembered or imagined, left behind by betrayal, violence and the war dead in Vittorio Veneto. These are part of the material culture of the living, "used to make representations about the past, the present and the future" (Renshaw, 2011: 27) – but they also constitute the ghosts of the dead, haunting city streets with reproachful grimness.

Devising the maps

In order to conceptualise the possibility and potential of the war maps, I first consulted with local experts, among whom was Pier Paolo Brescacin. Thanks to these scholars' encouragement and interest in the walking/mapping project, I visited the local and regional archives where I found out all I could about documented acts of violence, killings and ambushes in the town between 1943 and 1945. Using the spatial information from my archive notes and Brescacin's books, I then roughly mapped out the towns and localities that had been the setting of violent acts or potentially violent acts (in our case, failed ambushes). Slowly I formed an idea of what I wanted to ask participants during fieldwork. The maps were intended to situate places veterans and eyewitnesses considered important. I walked around with participants to allow the freest possible scope of movement in their remembrance and recognition. By deciding not to ask the co-researchers about any place in particular, I was hoping their footsteps and their comments would populate the memory/imagination maps themselves and provide a comparison with my own research notes and impressions.

With my co-researchers and colleagues, we chose to embody the stories, the memories and the emotions in a format which can be, if not quantified and catalogued, at least condensed in a 'practical' form. So we made maps, gathering together the main episodes of the civil war in the territory of Vittorio Veneto and surroundings. We did not wish to just contextualise a world of personal histories and memories, but above all to convey the sense of how deeply entrenched, implicated and interconnected memories, people and place can be. A format like a map can also summarise and contextualise facts and their protagonists in a way that might make sense to non-academics.

In mid-May 2013, I advertised for volunteers among a still large pool of local war veterans and individuals who had been alive during the war. I placed an advertisement in a local newspaper and recruited mostly through word of mouth, rallying up 21 people interested in sharing their memories and impressions of various wartime locations in greater Vittorio Veneto. Participants were aged between 19 and 87. Three of the volunteers had been active combatants in the civil war: 86-year-old FP had served in Mussolini's Fascist Guardia Nazionale Repubblicana (GNR henceforth) while 92-year-old Renato P. and 88-year-old Aldo A. were former partisans. Ten more individuals of both sexes had been teenagers and children during the war. Crucially, the wording I used to recruit these three older gentlemen, in particular, was quite different: they were contacted by a trusted colleague and asked about "wartime" memories, not about "civil war" memories; the fact of whether there was a civil war (or not) in the town depends on an individual's political leanings.

I asked my co-researchers to lead the way to the wartime places they considered important. I wanted to give participants the freest possible scope of movement in their remembrance and recognition. After dividing the volunteers into two groups, I asked participants how much they knew of the events that took place in the town during the civil war. My purpose, in following the local residents and the veterans, was to perhaps glimpse these intangible memories and weave them into the same complex story of fabrications, make-believe and celebration that permeates the city streets of the greater Vittorio Veneto area. I was aware that many memories manifested in the landscape leave little if any physical trace. The two maps that follow draw on oral history and archival data backed up by ethnographic fieldwork in Vittorio Veneto. The assembled information serves to visualise traces of wartime memory in the vernacular. Street names, rumours, memorials and private acts of revenge converge in an entanglement of memory and imagination at various levels, which then shape place (Muzaini, 2016). Among the locales on the maps are the three sites where the violence-redolent stories here were enacted either in the past or in the present.

> Just below there was a German command, and they often searched civilians in public places. There was much reason to be so afraid, you know. The Oderzo X Mas[4] arrived on a truck, they stopped in the square [Piazza Salsa] and they tell you: open your bag. Once one of that lot, he . . . he teased me with the dagger, asking if there were bombs inside my bag, to which I replied that he was welcome to look inside but there was only milk, as I was coming back from the dairy. Crazy stuff.[5]

Three of the infamous Fascist X Mas battalions landed on Vittorio Veneto on 30 October 1944 and set up their HQs in local schools. Battalion Barbarigo was headquartered at the Enrico Gotti barracks and in a wing of high school Lucio Classical Marcantonio Flaminio. Battalion Avalanche was lodged at the primary school Francesco Crispi; finally, Artillery Group San Giorgio settled in the Dante Alighieri College (Brescacin, 2012: 147–148). The maps that follow contain places that left no material traces: for example, the Fascist X Mas (Decima Mas) torture rooms in the basement of the Francesco Crispi primary school and the nearby Marcantonio Flaminio (Classics high school). These sites of atrocities live on in the memory of the older citizens, but their traces had to be thoroughly purged to redeem the buildings – understandably. Another invisible memory-place is the site which two former Fascist sympathisers had identified as the spot where spy Giobatta Brescacin had been ambushed by the "treacherous partisans".[6] This is the setting of the first place-story that follows. It is an anomalous lieux de memoire, fuelled not by mnemonic canons and stylised remembrance patterns, but rather by guesswork and curiosity about the dynamics of the event.

Most interestingly, the most striking example of discordant memories/warped imaginaries occurred at a site outside Vittorio Veneto, in the neighbouring town of Conegliano. Here, the medieval castle had been used by the Fascists to imprison and torture members of the local Resistance (De Nardi, 2017). Except that the local imaginations, which had dubbed the castle's towers "bloody towers" or "the butchery bastions", had populated them with bloodthirsty Germans. The imagination had willingly or unwillingly substituted the home-grown Italian Fascists with a much more palatable foreign enemy: Hitler's minions. I was struck by the attempt to banish the most horrible among local wartime memories to another town: as if by pushing the epicentre of monstrous violence away from themselves, and locating it elsewhere, Vittorio Veneto's mnemonic community was washing its hands of its own internal aberrations and bloodletting. The absence of similarly gruesome locales in Vittorio Veneto proper marked the town as bad and ugly, yes, but not "as bad as Conegliano".[7]

> This is the partisan Possamai, battle name Irma, captured and tortured at the Castle of Conegliano. They tortured and raped her so badly that she could not have children after the war. The castle was a seat of the SS where they interrogated and butchered people.[8]

As part of an imaginative geography of the civil war, this insight was off-scale: it perverted historical accuracy for the benefit of a less dreadful sense of place. The Castle of Conegliano represents, to the inhabitants of the nearby town of Vittorio Veneto, a lie viewed from a safe distance: in other words, what happened elsewhere can be reinvented to suit the requirements of the collective imaginary (see also Cole, 2015). Similarly, the foreign German militias had discreetly and conveniently replaced the home-grown Italian Fascists as the butchers. If more people remembered the fratricidal torture taking place at this contemporary locus amoenus, we may be sure that not so many would flock to it for recreation on a summer afternoon. Different (selective) imaginaries have shielded the real-world castle from its still more sinister legacy.

After fieldwork, I attempted to use some of the 'data' collected in the archive and layer it over the emotional responses collected from my participants. The discrepancies were more interesting than the confirmation of the existence of 'special places': the added layers of the imagination and the make-believe seemed the most apt to answer my questions. The imagination and the senses play a vital role in the mnemonic encounter in the vernacular of town and city. Memory is alive with images, emotions and "fantasies shifting yet real" (Kavanagh, 2000: 9). It is part of an assemblage that coheres through human and more-than-human intra-actions (*sensu* Barad) in the urban fabric. The memories and the stories told and untold build a geography of the past-in-the-present, multiple hauntings.

When they navigate the topography of the modern town of Vittorio Veneto, those old enough to remember may superimpose or slip into alternate places in memory and in the imagination, reliving or reimagining what went on (Figure 4.1). Younger individuals, on the other hand, may start to imaginatively feel close to the stories and people they heard about or read about in street signs, plaques and recollections, positioning themselves in a historical 'a-whereness' of the fratricidal war; they may perceive its potent affects vicariously (see also Light and Young, 2016). And overall, people need to feel 'at home' in this town to be able to fully uproot its secret past and to feel comfortable enough to express their views (Figure 4.2).

In the following micro-sections, I present the English translation of three memories-in-place. These vignettes arose during the memory walk/mapping experiment and stuck with me as the most poignant mementos of the power of memory *in place*.

Three emplaced memories

Story 1. Frustrated expectations

Giobatta Brescacin was a local Fascist spy. Everyone knew that he was, and he made no mystery of it. He loved delivering blacklists of anti-Fascist citizens of Vittorio Veneto to the local Fascist HQ. He thrived on betrayal and violence. So, the partisans decided to take him out. The safest way to operate was at night, with the cover of darkness. . . . The Resistance fighters belonging to the Communist Garibaldi brigade organised an ambush for Brescacin on the night of 19 April 1945. Right here where we stand, right now. At this crossroads. The partisans were waiting for him to go home. Only, a neighbour must have alerted him to something untoward and he slipped through the net. Brescacin did not materialise where he was supposed to, that night. Well, two nights later, the Resistance tried again. This time Brescacin approached but something, somehow, led him to turn back and flee.[9]

I have long been interested in the elusive, night geographies of the civil war. I was aware that the partisan war was a guerrilla conflict where both sides engaged in tactics and expedients were often dictated by opportunistic events and always

Figure 4.1 The 'walking memory tour' map of the centre of Vittorio Veneto (left and right)

Source: Copyright Sarah De Nardi

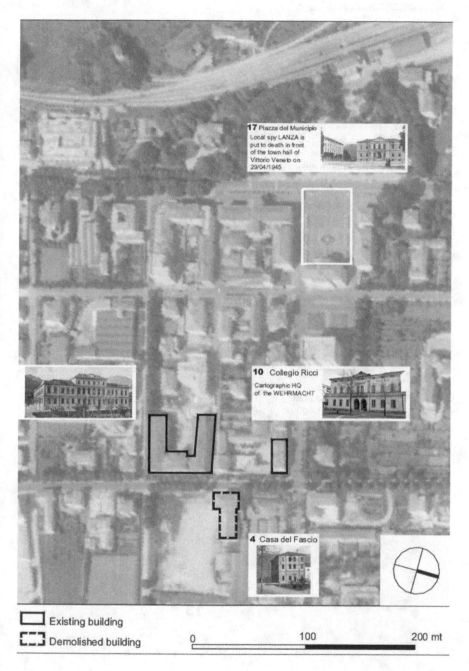

17 Piazza del Municipio

Local spy LANZA is put to death in front of the town hall of Vittorio Veneto on 29/04/1945

10 Collegio Ricci

Cartographic HQ of the WEHRMACHT

4 Casa del Fascio

☐ Existing building

☐ Demolished building

0   100   200 mt

Figure 4.1 (Continued)

MAP OF GENERAL AREA

— Main roads

■ Episodes of partisan activity

⟷ Partisan actions

■ Episodes or places of fascist activity

CANSIGLIO plateau
1000 mt a.s.l.

Rugged and wooded sloping
area suitable for partisan actions,
enemy observation and shelter

Monte PIZZOC
1500 mt a.s.l.

VITTORIO V. 138 mt a.s.l.

1	HQ GNR at Fratte Carron	13	Querry of Carpesica
2	Friga at SARMEDE	14	Fadalto
1a	Albergo San Marco	15	Borgo Fadalto reprisal
1b	Bus de la Lum	16	Osigo reprisal
3	Cavril hill	18	Cordignano partisan attack
5	Villa Vianello at Cozzuolo	19	Cordignano, riverbank
8	Borgo Gava	20	Canonica of Montaner
9	Pinidello, Meschio riverbank	23	Brescacin is wounded
11	Tarzo cemetery	25	SERRAVALLE La Giraffa

▪ ▪ ▪ ▪ ➤ Main movements of German army

0 0,5 1 1,5 Km

Figure 4.2 The territory of Vittorio Veneto

Source: Copyright Sarah De Nardi

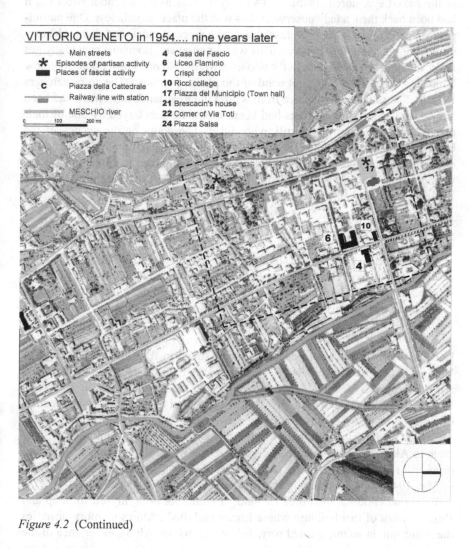

VITTORIO VENETO in 1954.... nine years later

———	Main streets	**4**	Casa del Fascio
✱	Episodes of partisan activity	**6**	Liceo Flaminio
▬	Places of fascist activity	**7**	Crispi school
C	Piazza della Cattedrale	**10**	Ricci college
———	Railway line with station	**17**	Piazza del Municipio (Town hall)
▬		**21**	Brescacin's house
	MESCHIO river	**22**	Corner of Via Toti
0 100 200 mt		**24**	Piazza Salsa

Figure 4.2 (Continued)

aligned with the 'affordances' in their environment (De Nardi, 2016). That day, on the walk, we learned that in the midst of the perilous sly nightscapes of the guerrilla, members of the local Resistance has been hiding in wait for a man who never materialised. The address where two unsuccessful ambushes awaited Brescacin on 19 and 21 April 1944 was the corner of Via Toti and Via Del Fante. Invisible to the naked eye, unremarkable in its everydayness as a residential street (as it had been back then, a lady observed), this was the place nonetheless. One participant snapped away at the spot, fascinated by a piece of cloak-and-dagger history unwrapping itself before his eyes. This was a non-event. Someone eventually got Brescacin and wounded him, but it was afterwards. No one knows precisely how Giobatta Brescacin, the spy, got wind of the planned ambush. The co-researchers felt sure that reliable intelligence from within the Resistance movement had been passed on to him. The partisans had been, most likely, betrayed by one of their own – and we do not know, still, who did it. The doubt on the agency of the betrayal remained on us, the walking party, even after we returned to base. Someone, among the good people of Vittorio Veneto, had talked too much and to the wrong people. And we wondered about what had told Brescacin to turn away the second time – a certain feeling in the still evening air?

Story 2. Unknown bodies

> I was deeply shaken by the sight of the execution of a spy in front of the town hall of Vittorio Veneto, a man called Lanza. I remember all his hair stood on end, and he was terrified. He . . ., he was clinging to the priest who had been comforting him, screaming, "Don't go! Soon as you go, they will kill me". Ghastly. I've always been a sensitive type, and always rejected violence.[10]

Local spy Lanza was put to death by firing squad in front of the town hall of Vittorio Veneto on 29 April 1945. Historian Pier Paolo Brescacin has identified Alfredo Lanza as Nello De Vido, a local spy to the Germans (2014: 109–110). His intelligence and cooperation with the enemy led to the arrest and execution of no fewer than 30 Resistance activists.[11] We know that now, that is – who Lanza was, what he did. That information had not been available in the immediate aftermath to Aldo – too unpalatable to divulge in the weeks and months of post-war grief and reconstruction. Street parties after the town's liberation by the Allies (1 May 1945) took the cheering crowds across that fateful square.

Where was Aldo at the time? He said, 'I could not bring myself to jump and shout in front of that building where Lanza had died'. Aldo's memory unsettles the status quo in an unexpected way. Taken as a whole, Aldo's experience of the war had been positive, and he took great pride in what the Resistance and his unit accomplished – and yet, there was this one incident. Simply talking about it was not enough – he had to show us where it happened. And this act, physically revisiting the place, was important to him and to his attempt to make peace with his conscience. The mapping, the walking to the spot – these acts helped him. That is all that I could possibly hope for – for Aldo to 'face his demons' and overcome them, supported by the caring community of other older citizens and sympathetic younger people. Most importantly, the showing of the location of the incident served to bring the rest of us closer to his experience so we too could partake in the rememory of the trauma he had undergone. We felt closer to his experiential and mnemonic world. This is how participatory fieldwork works at its best: when participants feel a 'something' and share an experience together.

Story 3. This one would not go home: the killing of Baldini

This is where he was made to stand, in front of the church, those cowards. It was 30 April 1945, and the civil war was not over, oh no. Blood was flowing by the bucketload. The Communists made Baldini stand there, the priest comforted him and gave him, you know, his last rites. I can only imagine how scared he was, so far away from home, never to see his loved ones again.[12]

Piazza San Michele or Piazza Salsa was to be the only site where we did not encounter a place-event of violence through eyewitness accounts but through another potent material culture item: a photograph. On a crisp April day in 2014 we arrived at a little unprepossessing square, led to the site by the former Blackshirt FP. The elderly man was visibly brimming over with emotion as he started fumbling in his case. That is when our day took an unexpected, interesting turn.

FP extracted from his bag two A3 poster-size prints he had had made from originals held in the local Resistance institute archives (ISREV). After showing the enlargements to all participants, FP started telling the story of the execution of GNR lieutenant Armando Baldini on 30 April 1945: days after the liberation of most of Italy by the Allies, aided by local Resistance forces. The decision to kill Baldini on that April day is still contested (De Nardi, 2016), his demise seen by many as punishment of a token Blackshirt-wearing officer and ultimately as a gesture of revenge towards the Fascist foe who had brought disarray to the town. Moreover, Baldini was not a local, coming as he did from a region in central Italy: an outsider and 'enemy' officer among home-grown anti-Fascists. After outlining the known facts of Baldini's death, the veteran began using the photographs like a map, pointing out different angles and trajectories in the square. In his slow and deliberate actions, the elderly man was embodying the ordeal and demise of the uniformed officer shot by the partisans. The square had been chosen as a gesture to avenge young local partisan Giuseppe Castelli, shot in front of the church on 6 February 1945 (Brescacin, 2014: 32–34). Baldini was executed on the exact same spot as the young local partisan. In a chilling chain of events, Baldini was put to death exactly one day after Nello De Vido aka 'Lanza' – the spy whose death had so troubled Aldo – had been executed on the main square. The last flicker of order and discipline had all but dissipated into chaos in Vittorio Veneto, and there was no end in sight.

The two squares (San Michele di Salsa and the town hall) differ in scale and 'feel', being situated in different locations in the centre of town, but they are both reminiscent of deaths in various ways. The Resistance did not forgive and forget,

Figure 4.3 The killing of Armando Baldini, lieutenant in the Guardia Nazionale Repubblicana
Source: Copyright ISREV Vittorio Veneto (with permission)

said one participant who wishes to remain anonymous. FP, by reliving another's ordeal, rehearsed his proud Fascist identity and legitimised the struggle of those who, like himself, had believed in the cause of the fatherland and trusted the duce, Benito Mussolini. The participants in the town walk and the 're-enactment' that day placed themselves and their living bodies in a phantomic space that had been occupied by living and dying individuals all those years ago. A performance of memory ensued, aided by archival sources, memory and the imagination of the participants. In a forthcoming journal article, I more exhaustively explore the performative nature of the 'placing' of the executions in situ. Its significance in relation to the visualisation exercise is that on the map we had, side by side, two dead men of different political affiliations, two victims of the civil war from the opponent sides, united in the evocation of their lives and their names by cotemporary Italians who wished to remember. Had we not been in place, on the Piazza Salsa, right where these unfortunates were killed, their stories would have lacked the potency of rememory that they were able to evoke for us on the day. I think we were grateful to those custodians of memory who could face the facts and relive with us those fateful days.

Concluding remarks

"Memory may be thought of as something beyond a sanctioned ocular association, [. . .] through a visceral shudder, *felt at the heritage space itself*" (Tolia Kelly, Waterton, & Watson, 2015: iii).[13] The alleged unreliability of oral history as a materialisation of memory constitutes its "strength" (Rouverol, 1999), and the subjectivity of memory provides insight into "the meanings of historical experience" (Cole, 2015: 19). In the case of wartime memory in Vittorio Veneto, the subjective colouring of the imagination had revealed hitherto unexplored geographies of the civil war which I could use to build an emotional and affective map of the town's remembrance and engagement with its unpalatable past. During fieldwork, we made up a mobile and dynamic tapestry of remembrance and imagined loci of wartime action as we went along. Where memory failed, the imagination took over. We made and learned about place on our feet, shuffling about, murmuring in the presence of particularly vicious sites of carnage. Walking and physically being in place constitute an important aspect of how the places are experienced or remembered today. Walking discloses and facilitates particular experiential modes in the city, alternating real and imagined places through physical immersion, memory, appreciation and learning of things familiar and unfamiliar. Fabrications, gossip, hearsay, legends and actual memories intermingle in the entanglements of memory and the imagined.

The most moving fieldwork moment was when former partisan Aldo A. admitted that he still associated the space in front of the town hall with Lanza's execution. Six months after the end of the Italian war, when the dust was only just beginning to settle on wartime violence, Aldo had done some rooting around, trying to find out about the man Lanza. Despite his research and what became a personal quest for clarity, Aldo never found out much about the man's life. Lanza's pathetic death was all he ever knew of him. And yet, to this day, "when I cross the square in front of the town hall I see that poor devil, as vividly as if it were yesterday". The ghost of the spy shot by the partisans haunts the memory of one of the partisans in the town – one who was once a member of the group that history had elected as the good guys. According to the shared ingroup memory of Aldo's veteran circle, the partisans were entitled to bring the spy to justice. Few would argue with this point, including Aldo himself – and yet the anguished memory of the act still troubles today.

The haunting of the memory of the execution is strong in Aldo's imagination. Even when Aldo tried to give a face, a name, to the haunting, he could not find out who the victim had been. The town square is adumbrated by the memory of that day which clings, stubbornly, to the now bloodless flagstones. For Aldo, his experience entails a simultaneous process of presenting and distancing. This "doubling" creates a residue. Some highly unorthodox memories have mainly been rehearsed through the bodies and relics of the victims of violence. My own grandfather, a prominent partisan in the Communist Garibaldi brigade, admitted to a fellow Resistance fighter:

> Sometimes we got it wrong. We were human beings, caught up in a nightmare of violence and assaults, and shit-scared at that.[14]

Any civil war is messy and ugly. In Vittorio Veneto, the collective and individual imaginary and the remembrance of the civil war have not entered into a fruitful dialogue. Too much hearsay, too much prejudice, too much make-believe exists for that to happen. Further, our encounter with violence tends to be profoundly personal and "fundamentally linked to processes of self-identity and the politics of personhood" (Nordstrom, 1997: 4); the site of one man's salvation might be the site of another's violent demise. The ideas of presencing, distancing and absent pasts are central to my argument here. Thinking back to the haunting memory of Lanza's execution, we might say that the past is present in the partisan's memory *and* imagination. The town hall square contains so many more meanings and imaginations, however: mapping Aldo's sad memory of the killing is just an anomalous layer over the site of the final triumph of the Resistance at the end of the war: the blood washed away by patriotic pride in a noble endeavour.

The loss of lives and the loss of the memory of the dead can only be processed, decanted and resolved across generations (see also Barbour, 2016). Blum and Secor maintain that trauma is not a single event which can be pinpointed in time and space, but rather a "topological constellation. The topological structure of trauma . . . is at once the cause and effect of psychical repetition phenomena" (2014: 105). The idea of a traumatic constellation brings us closer to the theme of networked memories and networked affects. It has become evident, in the previous chapters, that place and memory can unite or separate entities in time and place but also members of the same community if these members do not feel like they are inhabiting the same space, psychologically, emotionally or politically.

Taken as a whole, the mapping exercises my co-researchers and I have grown to enjoy so much represent (perhaps) an attempt to bring together all these different constellations of place-meaning and to consolidate them even if only for a moment, on a piece of paper, as 'their place'. A map could help us work through some memories, impressions and postmemories that are hard to talk about. The maps of the community in Vittorio Veneto, on the other hand, conjured up the troubled and troublesome memory of a non-absent past (Domanska, 2006). In-between, shifting and ambiguous shades of collaboration, collusion, culpability and violence that had been left largely unstirred, living at the edge of speech, confined to the intimate realm of the family history, whispered behind closed doors, were made visible. Ghostly photographs were re-enacted in present spaces, a last act of respect for the 'enemy' dead (as in FP's moving performance).

I would compare the act of mapping emotions and imaginations in the vernacular to an archaeological investigation, in a way that disrupts, destroys, invades and peels back layers of buried experience but also creates knowledge in the community (Perkins, 2004). Mapping may address blank spots (Crampton, 2010). It may unlock "the silent and painful lacunae in our understanding of recent experience. [. . .] It almost invariably goes to the heart, or more accurately, the painful nerves and tensions of experience that would disrupt and/or challenge the dominant voices structuring our experiences" (Buchli and Lucas, 2001: 15). Moreover, recognising the blind spots in memory and their place in the affective and mnemonic spheres could serve to exorcise and lay a few ghosts to rest. Perhaps mapping is a way of reckoning with and coming to terms with a painful heritage (Aksam, 2007; Sather-Wagstaff, 2011). Although the present case studies have contextualised the wartime memories of an Italian town, the deep mapping exercise could perhaps be beneficial to other, equally troubled, mnemonic communities across the world.

Summary points

- Having two different scale maps of the area made more sense to the co-researchers than sticking to a large-scale one that would necessarily cram lots of information: we decided to make a town centre–specific one outlining micro-geographies of war events and a wider one documenting the movements of German and Fascist armed forces in the wider territory
- The map of the town centre was the outcome of walks and site-specific mnemonic exercises and as such encompasses sequential, summative and point-of-view-specific referents relating to events and places in the fabric of the town
- The wider territorial map was made by our team of co-researchers following clues in history interviews and archival searches, as well as a memory walk in the outlying areas
- Present and absent elements of the townscape blended together: present and past uses of buildings during and after the war appear together on the map
- Intangible and tangible material culture found space side by side on the map
- Some co-researchers had not personally lived the events in the town or territory due to their age; at the same time, they were brought up with an affectual postmemory of the events that, in their perception, haunts the town

Afterword

Despite some of the stories of violence and revenge embedded in this chapter, on an affective plane, my loyalties and sympathies go out to the Resistance movement. The struggle and sacrifice of the partisans are officially celebrated every 25 April in Italy as the democratic rebirth of the country. I will never cease to be proud of my grandad and his peers – for better and for worse.

Notes

1 Renato P., interview, 24 September 2011.
2 My late grandfather Domenico was a prominent local partisan; at the same time, my father's uncle Franco was a draft dodger and abscondee. Last, my maternal grandmother's brother Ivo had been a bona fide Fascist enlisted in Mussolini's army who was killed in a partisan ambush in Lombardy, far from home, in 1944.
3 SOE was the British intelligence corps sent to aid and train Resistance groups.
4 Special Navy elite group of the Fascist armed forces. They were infamous for their violence.
5 Anna G., interview, 27 August 2013.
6 BD, interview, 3 March 2015.
7 Giovanna F., interview, 24 June 2014. Domenico F. (Interview with Serena Conte and Chiara Strazzer 2002, *Intervista a Domenico F., ex – Partigiano, e al Signor Bruno A., ex – Fascista.* Unpublished) described the Castle of Conegliano as follows: "It was the X MAS's Intelligence gathering headquarters and a place of torture and killing of partisans and civilians. During interrogation the Black Shirts tortured prisoners, tearing out fingernails and cutting an X on their flesh, on the breasts and backs of captured women and men. The Castello is truly a place of pitch-black memory".
8 Anna G., interview, 25 August 2013.
9 Pier Paolo Brescacin, interview, 25 October 2014.
10 Aldo A., interview, 22 December 2012.
11 Pier Paolo Brescacin, personal communication, 25 November 2014.
12 FP, interview, 1 April 2014.
13 Tolia-Kelly, Waterton and Watson 2015, p. iii, my emphasis.
14 Letter to Raimondo "Chirurgo" Lacchin, 25 April 1994. Private Favero archive.

5 Unlearning the body
Liminal spaces, abject corpses
and the historical imagination

Foreword

Is it not true that the only dead who return are those whom one has buried too quickly and too deeply, without paying them the necessary respects, and that remorse testifies less to an excess of memory than to a powerlessness or to a failure in the working through of a memory?

(Deleuze, 1994: 15)

Does remorse, or the lack thereof, testify to an excess or a failure of working through of a memory? In this chapter I argue that over the years, narratives of the events of the Italian civil war (1943–1945) have been created, re-enacted and relived based on Othering practices shaped by processes of infrahumanisation. I look for the space of the dead, the Deleuzian dead "whom one has buried too quickly and too deeply" (Deleuze, 1994: 15); I excavate the ambiguous space the corpses exhumed from a mass grave occupy in the historical imagination. The bodies of (largely) unidentified wartime victims embed their own geography of silence and forgetfulness in a mass grave. In bearing witness to the unearthing and reburial of these problematic human remains, we experience an erasure of accepted ideas of the body.

Il Messaggero, 25 July 1966

Five hundred people '*infoibati*' down a vertical cave in the Cansiglio plain

> Four young speleologists [. . .] claim to have found hundreds and hundreds of dead civilians, Germans and Italian Republican soldiers in an hourglass-shaped cave reaching a depth of 285 meters. They [claim] the bodies would have been thrown in during the stormy final phase of the war.
>
> The four explorers, the first to descend into the cave, had allegedly inter-rogated the population of the area on their knowledge of the cave. Bus de la Lum or Lum, translated in Italian as Hole of the Light or of the Moon, has long been known to the locals due to the periodic occurrence of will o' the wisp in its interior.

The cavers interviewed people who "know the story"; they heard from a person who, under the influence of alcohol, allegedly confessed to have personally thrown soldiers into the cave; and they collected testimonies of numerous farmers who claim to have seen the aged relatives of the 'disappeared' praying near the Bus de la Lum, even today. At 180 m below the entrance, there are reports of the recovery [by whom?] of plenty of corpses, but these are putrefied. [. . .] Ultimately this is a series of stories that, if taken seriously, leads to the following conclusion: in 1945, over a hundred people between soldiers and civilians were disposed of down the cave [!!] (Figure 5.1).

Il Messaggero, 3 August 1966

A fanciful tale debunked: no corpses in Bus de la Lum

At our explicit request 15 speleologists from Friuli and Trieste carried out an exploration of Bus de la Lum on Pian del Cansiglio where, according to rumours, the bodies of five hundred people thrown into the vertical cave during the last war lie unburied.

Figure 5.1 No *infoibati*: a newspaper cutting from 1966

At 180 meters from the entrance of the Bus de la Lum, at the depth of around 225 meters, the speleologists found an obstruction, a pebble and crushed stone platform which blocks the way to further descent. On this platform with a surface of 30 square meters, there is no sign either of skeletons, or of objects. [. . .] Readers will remember that days ago, in our report about an alleged mass disposal of bodies inside Bus de la Lum, we expressed some doubt. However, some cavers from the local rescue centre had told us about the recovery of corpses encased in ice blocks, with their military uniforms still intact [whose? BlackShirts? Germans?]

[. . .] Moreover, the population of neighbouring villages in Pian del Cansiglio has nurtured for years the idea of *infoibamento* in the Bus de la Lum, in relation to the possible fate of numerous people who disappeared during the murky last phase of the war.

>>>>> Fast forward to 1988.

La Squilla, November 1988

A key figure in the esoteric fascination around the Bus de la Lum was Don Corinno Mares, parish priest of Tambre (Belluno). It was he [see photograph below] who erected the "Silentes loquimur" cross at the mouth of the cave on August 29, 1987 [as reported in Tambre's La Squilla newspaper, November 1988, under the piece "Cansiglio: A cross at the 'Bus de la Lum'"]. From then on, every year on that day, the priest has celebrated a Mass for the "hundreds of civilians" once buried deep in the cavity. A former partisan sued him for slander against the Resistance and the case went to court.

(http://dati.camera.it/ocd/aic.rdf/aic4_01539_11)

La Tribuna, 11 March 1989

Harsh rebuke of the ex-partisans: "no atrocities on Pian Cansiglio"

Those murder victims represent one of Italy's darkest histories. [. . .]

The partisans, who lived through the twenty harsh months of the Resistance, deny any atrocities while admitting that a dozen Fascists, plus three captured spies had been tried, found guilty, executed and made to disappear into the cave when already dead.

According to the former partisans: "There are no forgotten massacres, only specific facts which we can explain". Thus: No Fascist was thrown alive into the "Bus de la Lum" (Figures 5.2 and 5.3).

Remembering through an infrahumanisation optics

The theoretical framework for my thinking in this particular chapter borrows from theories of infrahumanisation in social psychology – see the previous chapter for

Figure 5.2 The Trieste speleologists who explored Bus de La Lum in the 1960s

Source: Copyright ISREV (with permission)

an analysis of the infrahumanisation agencies behind wartime killings. To summarise briefly, researchers argue that we, as human beings and social groups, tend to discriminate against those perceived to be 'different' from us and belonging to an 'outgroup' (Leyens et al., 2000). Outgroup members, social psychologists believe, are not attributed the ability to feel the full spectrum of human emotions. (Leyens, 2009). For instance, primary emotions such as pain, fear and joy are ascribed to all human beings, but other, subtler nuances of emotion such as pride, regret and compassion are attributed to some, but not all. Infrahumanisation theory presumes that in order to behave humanely towards an individual or social group, we must acknowledge their full humanity or even their capability to feel, exist and act as we do. The moment we perceive a group or individual as inferior in emotional terms, we may feel entitled to judge them, criticise them; we may

Figure 5.3 The speleologists

feel entitled or allowed to dismiss their rights and do things to them we would not do to our 'perceived' peers.

This ambiguous othering practice has everything to do with identity. Writing from a sociological perspective, Burkitt posits: "self-identity is still constructed in their relations to others in many important ways, in families, through friendships, workplace relations, or indeed in any forum or medium through which individuals come to identify or *disidentify* with each other" (2012: 460, my emphasis). Disidentification may be an inner, almost intimate manifestation of the workings of infrahumanisation that manifests with racism, discrimination, hate crime. The seed of this powerful yet destructive mode of being and doing lies in the suspicion and exclusion of an Other. If discrimination and infrahumanisation of specific groups in memory based on their ideological identity, as it were, impinges on identity-making, then discriminating against groups and individuals who were perceived as 'infrahuman' during the war may be a reason for the deep fractures existing in many contemporary societies across Europe today. Might this have been the case in Italy, Chile, Argentina and post-Franco Spain? The implications of infrahumanisation theories for the study and understanding of the European war and global anti-Fascist regimes are far-reaching. The tropes of infrahumanisation abound in remembrance narratives: just by way of example, wartime memoirs are rife with bestial epithets directed by the partisans towards the Fascists and vice versa. There is also evidence of a plethora of Germans' dehumanising words for Italians, with the epithet of choice being *Schwein* – pig.

The questions I pose in this chapter stem from, and develop around, the following preoccupation: what if the historical imagination of Italians is so fundamentally hardwired to reject or negate the proper humanity of political enemies in the civil war that the very idea of the sanctity of the body appears genuinely redundant? What if the veteran Resistance fighters genuinely believed (or still do) that it was acceptable or even necessary to execute suspected spies, not out of a clear-cut conviction of their guilt, but in light of their marginality – their being 'not quite like us'? Emotion plays a key role in how we include or exclude persons and groups in our circles. Consequently, I argue that assimilation and negotiation of identities, personhood and responsibility depend on perceptual factors: that certain personhoods have been denied the right to live and the right to mourning. It follows that unmourned losses are spectral presences haunting the location of a grave site and the lives of those who encounter or come upon that place of unrest (see also Boss, 2010).

A place we cannot know

I flipped again through the 1960s newspaper cuttings about Bus de La Lum; they appeared somewhat surreal, their discourse unashamedly sensationalist: the bodies of the murdered enemies of the Resistance were embroiled in a fantastic story complete with gory details (*putrefaction*) in abundance. The media imagination had already despoiled these human remains of their humanity and dignity in order to glamourise violence. By contrast, *La Squilla*'s 1988 newspaper cutting was hagiographic and heart-rending – it decidedly painted Don Corinno the priest as a saint for rendering justice to the innocent victims of the bloodthirsty Communist partisans.

Across the cuttings, however, the nameless dead still occupy a liminal (*sensu* Turner, 1967) and marginalised space in physical terms (a hidden cave), in memory (who were these people?) and in ethical terms (did they deserve oblivion?).

What can be learned about these remains? The victims have now been respect-fully reburied, but not all names are known. "While, rationally, death is final and irrevocable, experientially it is not. The presence of the one who has died (or the beloved from whom one is separated) lives on in after-images, longings, intima-tions, dreams and reverie" (Byrne, 2013: 606). The insistent presence of the dead peeks out from the decaying newspapers and demands to be noticed. Or is this affectual outreach of the dead part of the imaginative process of research? At any rate, shuffling through these newspaper cuttings during my research was as demoralising as it was chilling. I inspected these documents twice. Once when I obtained the full records from my colleagues at the ISREV archives; the second time with my community focus group/mapping collaborators, whom I introduce in the following. The inspection of the cuttings and the headlines with the latter group proved to be even more upsetting, but it was a necessary step to ensure that we were all provided with the same information.

The first time I handled these news pieces and as my colleagues at ISREV pre-pared a set of photocopies for my own record, we commented on the seemingly unending feuds and bad memories tethered to Bus de La Lum. Here I unpack some of the strands of this conversation and other assorted ongoing reflections.

Bus de La Lum constitutes a story that encompasses many tales: all ugly, sad and ultimately desperate stories in which Italians turned against each other in a civil war and continue to harbour an obstinate separation in remembrance, over and over and again. For Barbour (2016: xxxcc), "Unmourned losses are projected forward into the lives of future generations and [. . .] can only be contained, processed or decontaminated, over a span of generations. There are traumas for which a single personality and a single lifetime do not give enough space or time for working through". Scholars and individuals interested in European war events are routinely fascinated by the ambiguity of the historical representation of what occurred in Italy between September 1943 and May 1945. Much post-war rhetoric – the non-existent postmemory of the victors and the Allies – has kept quiet on the civil war raging through the country alongside the thrust to liberate Italy from the Nazi occupation.

Ever since the end of the war, the left-wing veteran partisan association, the Associazione Nazionale Partigiani d'Italia, has acted as the key memory gate-keeper; it has performed a pedagogical and moralising role aiming for a demo-cratic education of the young through curriculum programmes and youth outreach activities (Clark, 1984). Countless ageing veterans have been going out to schools, talking about their wartime experience fighting Fascists and Germans, with the heart-warming pathos and righteous civic pride of individuals who know they have served the cause of justice, democracy and patriotism. On the other end of the scale, we have the outraged, rancorous, private gloom and mourning of the right-wing families and descendants of ex-Fascists and the lugubrious, angry nar-ratives of the neo-Fascists. These stories inhabit the same places, contending for attention.

But what about the cracks in between the worlds, the worlds of the anti-Fascist heroes and the ever more vocal Fascist veterans? In the aftermath of the Cold War and the fall of the Berlin Wall, the shape-shifting Italian Left was left to combat a postmemory made of accusations, recriminations and a retroactive rage towards, and fear of, the Communist other. Cold War Italian political debate was suddenly suspicious of the motives and the ideological drive behind the armed struggle of 1943–1945. Much mudslinging ensued among the moderate Centrist parties, the Left, the Right and the popular imagination. This inability to properly position the Resistance movement was the root cause of much of the storytelling around Bus de La Lum. We found that we could not define where memory ended and the morbid imaginary of a community began. Not that this ambiguity needs trouble the research: rather, this opacity can often enrich the framework of inquiry.

The issue with memory work in general – I will not go into the hefty debate on the fallibility of individual recollection or the usability of oral history as scholarly data – is that it often over-reaches and sometimes gets it wrong (Thompson, 1998; Cappelletto, 2003, Radstone, 2005; Hrobat, 2007). In the case of the Bus de La Lum, some of the people I spoke to and some of the accounts that people volunteered did not reflect the truth. Some misremembered, as shall be seen. These incongruent or alternative versions of events were fundamental to the creation of the mythology of this place, but my main cohort of co-researchers would not include them in their visualisation. They decided that this other mode of remembering did not belong in the story they wished to tell. Hovering at the margins of speech and map, these alternate viewpoints formed a spectral geography of the "disquietly unfamiliar" (Kohon, 2016: 13)

The Bus de La Lum memory work was hard going because it was so divisive. Sociology has long compared the world-making role of memory to building blocks of social identities and cultural enactments. Memory is a social binder, asserting and keeping groups together – or keeping them apart. For Mistzal, "the reconstruction of the past always depends on present-day identities and contexts" (2003: 14). Moreover, "the only past we can know is the one we shape by the questions we ask, and these questions are moulded by the social context we come from" (Kyvik, 2004: 87). So far so good. But there may well be cultural and psychological schemata influencing and shaping the past *and* the present experience of war and conflict (Sumartojo and Stevens, 2016). Social groups and mnemonic communities might harbour and nurture postmemory tenets that are unquestioningly accepted and assimilated. Halbwachs (1992, 1997), and Nora's (1984, 1989) seminal works on collective memory have certainly shaped the ways in which anthropologists, sociologists and historians think about the mechanisms and politics of remembrance. Halbwachs's initial premise is that every social group "develops a memory of its own past that highlights its unique identity" (Misztal, 2003: 51). But what, I ask, if we were to build our own histories and identities on such fallacies of perception? What impact would this have on the politics of memory?

Owning stories

"I own my story: it is mine and mine alone", claimed a veteran partisan who claimed to not remember anything about Bus de la Lum. "I won't let anyone use my story against me", argued another, asking her name to be removed from my database. For the purpose of this chapter and the wider subject matter of the book, I will refer here to the stories that can and should be visualised. These, admittedly, could be never-ending, ever shifting and never resolved, but they can and should be understood and faced. By retaining a mournful silence or speaking in hushed whispers about the war dead, social catharsis cannot be accomplished. Instead, taking ownership of stories, taking ownership of a present, can occur through representations of a past.

The overarching blame on the partisans for the alleged atrocities perpetrated at Bus de la Lum has taken on vastly mythic proportions. It is a snowball effect. Someone starts telling his or her own version of the story, and so it spreads. "As soon as a story is well known – and such is the case with most traditional and popular narratives, as well as with the national chronicles of the founding events of a given community – retelling takes the place of telling" (Pavone, 1991: 110). The story has deepened and contracted depending on who the storyteller is/was. The heritage of this site is an apparently incompatible, many-voiced assemblage. Some of the storytelling took on mythical, hagiographic dimensions. The Catholic collective consciousness has cast doubt on the full humanity and decency of individuals who expressed a certain political credo and enacted a different, more politicised agenda. Is a Communist not a human being, the same as a Fascist? Do competing Italian narratives of the war function the same way that Polish war memories were constructed – around myths, misunderstandings and tragic mistakes (see Drozdzewski, 2012)? And yet, stories were and are still formed on a regular basis, everyday legends and long-term inimical tropes – almost daily narratives of reciprocal otherness. Italians fighting, injuring, killing Italians, the same but not the same.

Let us look, then, for the space of the dead at Bus de la Lum. We know the identity of 11 victims; ten were Italian Guardia Nazionale Repubblicana soldiers: Ruggero Ciruzzi, Riccardo Frare, Giovanni Brunello, Giovanni Birghin, Elio Romeo, Antonio Righetto, Antonio Brusadin, Giuseppe Boscarin, Sante Sperandio and Giovanni Rizzo. We know the name of the only female victim, 36-year-old Marianna "Nella" Dal Bo De Pieri, who was an informant to the Fascists and Germans and was executed by the partisans on 9 September 1944 (Brescacin, 2014). In total, 38 corpses were found, most of whom were in an incomplete state. The remaining exhumed victims are still awaiting identification.

The Cansiglio forest covers the uplands of a Karst massif that forms a mountainous promontory from the Dolomite massif that juts southwards into the Veneto piedmont and Po plain. As a high limestone ridge, this area has several high plateaus interspersed with vast areas of forest. Although very cold in winter and dry in summer, this forested area provided the shelter and safety for partisan bands that the flat expanses of the Veneto plain to the south and east could not. As such, this was a site of significant partisan activity against the Nazi-Fascist forces; as a result, at the southwestern corner of one of the high plateaus, the Pian Cansiglio, there is a cluster of memorial sites that illuminate the struggles over the memory of the Resistance, its actions and those of its Nazi-Fascist enemies.

In 2017 headlines from local newspapers on the Bus de la Lum reported the destruction of a tombstone by unknown vandals. The uncanniness of the story is given by the fact that many had not been aware of the presence of any such marker. Indeed, who had placed the stone, with what authority and on what historical basis? The plate seems to have appeared overnight, sprung up like any other forest mushroom, without an official installation, inauguration or acknowledgement. Three of my collaborators affirm that one day the plaque was suddenly there, but no one remembers when exactly (Figure 5.4). The plate is dated 2015, the 70th anniversary of the Liberation, and was installed by the families of 3,463 alleged military and civilian victims of the Nannetti partisan division.

Before . . .

The right-wing association Casa Pound (which has since formed a political party, Destra Pound) seems to have claimed the initiative. The figure of 3,463 victims appears in the book *The Ghosts of Cansiglio* by right-wing popular historian Antonio Serena, in which he enumerated the dead in the territories controlled by the Nannetti division as follows: 1,054 Repubblichini, 2,294 Germans and 115 "spies". For Italian political party Fiamma Tricolore, however, the figure amounts to 3,000, which encompasses all the individuals who were thrown to their death down the Bus de la Lum. A blog (Wuming Foundation, 2016) reports that local

Figure 5.4 The plaque to the Resistance dead

Source: Copyright Pier Paolo Brescacin (with permission)

Figure 5.5 The shattered plaque

Source: Photograph by Sergio De Nardi (with permission)

newspapers contextualised the act of destruction of the plaque in a wider context of vandalism at the other end of the political spectrum: the writers compare the destruction of the neo-Fascist plaque with the simultaneous damage to the official monument to the Resistance in Cansiglio by unknown Far-Right fringe groups (Figure 5.5).

And after! The shattering of the plaque speaks volumes.

The blog authors conclude with a disquieting question: what is the worst problem here? Is it the destruction of a plate or the placing of a false and inaccurate writing without any vague historical evidence or fact in the first place?

The non-silent majority

Silence as a strategy for forgetfulness or selective memory does not work in Italy. There have been attempts to erase unpopular facts and to hush talk of events, but the results have not been a dwindling of public discourse into whispered half-truths. Rather, the outcome of exposed hush politics was, and still is, akin to heated screaming matches in which everyone, and no one, is implicated. And at the same time, wilful ignorance comes to the rescue of those who prefer not to remember, engage or take sides in the ongoing debate. I begin here by listing three of the eight (public) reviews of Bus de la Lum on Google.[1]

Riccardo F.

February 2019 ★★★★★
 A place, a foiba *which is definitely worth a visit. Natural cavity misused in the past to make persons disappear.*[2]

Marilena D.

July 2018 ★★★★★
 Deep cavity of karst origin located in the Pian del Cansiglio woods. Years ago, scholars poured into the cavity colourful water that they had drawn from the Gorgazzo stream. There are several legends about the origin of the name, with the most common being the fact that shepherds saw a light coming out of the hole at night (probably due to the putrefaction of animals that had fallen in). During the last war, unfortunately, the bodies of prisoners or soldiers were thrown into the cavity. In fact, the remains of about thirty victims were found.

Luigi Pix

October 2017 ★★★★★
 A place of history and suffering.[3]

Daniele T.

December 2016 ★★★★★
 Lovely walk in the woods, where you will come across a hole (a Karst sinkhole) that looks quite scary! You can't see the bottom.[4]

Andrea T.

October 2016 ★★★★★
 Lest we forget. A place to visit at least once in a lifetime, if you happen to be in the area.[5]

Out of the nearly 85 reviewer-visitors of this site, only 2, Andrea T. and Mirko C., mention 'lest we forget'. Riccardo F. openly calls the site a *foiba*, Marilena acknowledges an accurate number of human remains, while Daniele T. chooses instead to list the geological peculiarities of the sinkhole, adding that the bottom is invisible and that the place evokes fear. An Ivan P. praises Bus de la Lum's potential for a family excursion.

I find it interesting that even in this small sample of reactions and responses to the place, the memory and perceptions of the site vary enormously. This disparity in perception is important in light of wider attitudes to this site. Nico Vascellari, an Italian experimental artist, even entitled a large-scale installation after the sinkhole. In a work that deconstructed space and nature and questioned sense of place in the forest, Vascellari constructed a vision of Bus de la Lum which is eerie and hellish and encompasses the 'bad memories' associated with the site and its strange, otherworldly natural beauty (Louise, 2016). But it's the people whose memory is tied to the place (or not) who take centre stage in the cultural and mnemonic construction of Bus de La Lum as a site. The victims have come to represent the site as a whole, their numbers oscillating wildly through time. The victims' stories, once confined to the private mourning of families, have entered the political arena. So, the tragic story of 36-year-old Nella 'Dal Bo' De Pieri aka Nella Dal Bo (Figure 5.6) has been appropriated by revisionist

Figure 5.6 Nella De Pieri: wife, mother – spy?
Source: Wikicommons

right-wing groups who published an article negating her collaboration with the Germans and celebrating her martyrdom (Figure 5.7) (in the now defunct Il Giornale d'Italia.

The hagiography of an exemplary victim – a woman, a mother – serves to sully the memory of the Resistance, whose members took her life. One might think that the killing of Italians by fellow Italians, even during an armed conflict, would lead to mourning and to post-mortem compassion towards the identities and relatives of the victims. But this is not what happened. As the grandchild of a Resistance fighter, I grew up knowing that the case study of the Bus de La Lum has always been our family's thorn in the side. It tarnishes our pride as Italians and as anti-Fascists. It was certainly a cause of distress for my grandfather Memi. What really happened at Bus de La Lum and why is just now being reconstructed. It is an ugly story of abjection, denial of humanity, a refusal to behave with piety and a negation of the humanity of the other's body. It is a 'double death'.

The historical imagination of places such as Bus de La Lum is grounded on the shaping and sharing of narratives about the war, as we are social beings enacting identities every day (Figures 5.8 and 5.9). We produce memory with members of our affectual ingroups, starting from a shared set of knowledges. So, if this context-specific dimension of remembrance is true, how do we re-enact stories of alien, alternate or Other materialities, bodies, identities and positionalities? The

Figure 5.7 Nella De Pieri is tried by partisans in the Cansiglio forest
Source: Wikicommons

Figure 5.8 The sign leading to the archaeological site of Bus de la Lum

Source: Photograph by Sarah De Nardi

Figure 5.9 The arrow in the middle of the forest

Source: Photograph by Sarah De Nardi

infrahumanisation of the other side in a social and political conflict also afforded a relative ease to perpetrate violence on the bodies of those who acted a certain way and to dispose of their remains accordingly.

Writing about the plight of Italian partisans, De Donà recalled a chain of events which has been corroborated by other testimonies in the same town, but his memory then fell into the trap of the imagined crimes of the Resistance in the Cansiglio (Ballone 2007: 102).

> During the war the Germans looked through the houses and arrested all men hiding there. They put them on trains bound for Germany. My brother-in-law, to save himself after jumping off the train, hid inside a barn for a long time. As he jumped off the train, he was wearing a Wehrmacht uniform, you see, and he then had a lucky escape. He was still wearing the Wehrmacht uniform when he bumped into a group of partisans. He was nearly thrown into the Bus de la Lum by the group of partisans who mistook him for a German soldier. He had simply masqueraded as a German to aid his escape. Only the intervention of one of his friends, a partisan who recognised him and vouched for him, saved him from an atrocious death.

Here, De Donà simply misremembered what his brother-in-law had told him. Maybe his brother-in-law told him this version of the story, in which case the latter was mistaken. The partisans would have never dared kill and dispose of German bodies into the sinkhole as that would have ensured an atrocious reprisal against the civilian population. The people they concealed – made away with – into the sinkhole *had to be Italians*. Elsewhere De Donà had already expressed a judgment about the fate of Fascists who were captured in the Cansiglio woods. "Fascists were violent against the Partisans, yes, on the other hands the partisans captured Fascists, took them to the Cansiglio forest and threw them in the vertical cave of Bus de la Lum" (Ballone, 2007: 100). He acknowledged the abhorrent nature of the civil war.

In my conversations with the local communities, the following ideas started to take shape. It might be that Fascists and spies, who were the active enemies of the cause of Italian liberation, had not been perceived as entirely human. The disembodiment of these persons from the physical reality of their death (then) and from memory (now) serves as a paradigm for othering and abjection of the enemy. There are precedents in Vietnam and Iraq for this behaviour on the part of the US army. Conflict sociologist McSorley (2012) argues that in order to facilitate soldiers' acts of fighting, wounding and killing, and to 'salve' their conscience, the adversary or enemy has to be rendered somehow dehumanised by way of conditioning and training. Then again, those corps had been deliberately induced to disregard the humanity of the enemy other, be it Viet Cong or Iraqi soldier. In Italy this kind of army-normalised conditioning does not occur: the self-styled guerrilla groups of Resistance fighters had not been trained to kill first, think later (Focardi, 2013). Most partisans were not soldiers, had not even served in the army due to their young age, and most of them (including the several women and girls who took up arms) had no idea of how to act in a conflict situation. Many found the idea of taking a life abominable.

The idea of the body central to the Catholic mentality, which is one of piety and honour for the dead, was suspended in the belief that the persons in question had relinquished their right to a decent burial the moment they adopted Fascist or Nazi beliefs. The Fascists were considered base, bestial and undignified by the anti-Fascists, in particular by the partisans themselves. One of my interviewees, GT, explained in no uncertain terms:

Q: Who behaved worse – the Fascists or the Germans?
G: They were both beasts, but the Fascists were sadists. If along the way I bumped into a Fascist, I would shoot, no questions asked. I would not spare them. The Germans, however, were less insidious – they killed without torture. The Fascists were sadistic, evil. A Fascist woman, infamous in my home village, tortured one of our messengers, a poor boy of 15 years, who looked even younger. After they captured him she proceeded to torture him to death. Some woman, huh. They did it very slowly, torturing him in subtle and sadistic ways. When they buried him, the kid was unrecognizable. At Liberation, we shot the bastard sow who butchered him by the church tower. The Fascists had a taste for blood, I tell you.

The necessity to hide the bodies of the Fascists and spies at Bus de La Lum was explained by the perpetrators as a tactical must – the Allies stationed nearby might have objected to the partisans' kangaroo courts and hushed trials. Here lies the painful crux of the matter. The questionable legitimacy of the killings at Bus de la Lum is like a chilling mirror image of the abnormal treatment of the dead. So what became of those bodies once the individuals were dead? Did these human remains become an abjection, unworthy of Christian burial in the imagination of the partisans? Did they become objects that could be disposed of in the most practical and logistically viable manner? And if so, how could these dead be imagined, evoked and represented?

Visualising a cultural-emotional 'minefield'

The Bus de La Lum sinkhole is a non-place; it is an ugly secret, a black hole, a menace to memory and forgetfulness alike. To the public international gaze, Bus de La Lum is hidden. It is invisible. Although it can be pinpointed through Google Maps, for instance, it cannot be seen underneath its lush evergreen forest roof. Google lists the sinkhole as a 'War memorial' – but to whom? Between 2009 and 2014 I carried out ethnographic fieldwork in Italy in order to investigate the world-making practices and the workings of identity and sense of place and emotion in the experience of the anti-Nazi and anti-Fascist Italian Resistance of 1943–1945. I chose to focus on the memory of participation in the national Resistance movement through interviews and participant observation in my native region of Veneto. Of course, one of my colleagues in Italy said that people will tell you what you want to hear. But is this true? Do people not actually share what they feel is the correct and accurate version of events (see also Portelli, 1997)? Well, perhaps neither is correct. What if the memory and postmemory of the Fascists and anti-Fascists is so fundamentally hardwired to reject or negate the proper humanity of their respective opponents in the civil war that the very idea of taking responsibility for or regretting wrongdoings and violent acts appears genuinely superfluous? What if the veteran Resistance fighters genuinely believe it was necessary to execute suspected spies because of their otherness and marginality – due to the spies not being 'like us'?

In other words, we, as social beings enacting identities in the everyday, produce and share narratives about the war with members of our affective ingroup, starting from a similar pool of knowledges and a shared materiality (Witcomb, 2012). After all, memory is grounded in objects, in place and in the senses; through the senses we make sense of the world around us and of persons and things interacting with us in the everyday. So, if memory is context-specific, how do we re-enact stories of anomalous or Other materialities, bodies, and positionalities? How do we frame stories about the war that automatically 'other' the outsiders? And what about emotion? Emotion, I would suggest, plays a key role in how we include or exclude persons and groups in our circles. Emotion is quintessentially social – socially produced, enacted and shared (Ahmed, 2004; Burkitt, 2012: 459). Emotions, however, inform how we perceive the other and how we assess their humanity (Leyens, 2009).

Bus de la Lum is an invisible place of mourning, a historical site (according to Google's label), a place uncannily present-absent to all but those who know about it and are able to reach it on foot from the ground level, walking through the trees. This is a negative heritage site as it embodies the material culture of "absent objects, absent people, and the commemoration of things that are no longer there" (Meskell, 2002: 562). How can we write and represent such a place? If the physical objects (the bodies) are no longer there, how do we encounter and experience its melancholy invisibility? For Paul Thompson, the memory process depends on that of perception: "In order to learn something, we have first to comprehend it" (1998: 129). In other words, we cannot properly remember something we have not assimilated, metabolised and made our own (see Fentress and Wickham, 1992). This applies to the 'roles' played by others and to the construction and negotiation of categories of person and social groups in the past – as well as the present day. In the case of Bus de la Lum, the refusal to accept events and to reconcile memories has produced a void space into which everyone can pour their resentment, but not much else.

My fieldwork in the Cansiglio region was an attempt to *com*prehend – literally to understand together with – the locals. So this is what we did. As can be imagined, the mood was sombre at the community meeting at Hotel San Marco in April 2016. No one looked like they particularly wanted to be there, and recent rain had muddied the whole forest and plateau, making walking around unpleasant and damp. So our group sat in the conference hall of the hotel, pondering the mood of the place we had gathered to 'discuss'. The declined invitation on the part of two leading Resistance leaders – due to ill health, which is highly likely given their age – meant that no one directly involved with the sinkhole in wartime was present. The co-researchers I worked with had been children in the 1940s or were descendants of the fighters and collaborators. The questions that follow were floated in the room during the event. My initial prompt was: what are we to do with this site? Who does it belong to? Who is commemorated there? Can we turn this experience into "a spur to transformation, to difference" (Grosz, 2008: 49)?

A woman (a schoolteacher in her 60s) pointed out that the site's main legacy is as a failure. When prompted to elaborate, she went on to say that for her, Bus de la Lum was a blatant failure of the Resistance – the most shocking stain in its often-immaculate track record. It is not my place to comment on the sentiments populating the room at that point. It would not be chivalrous to my co-researchers to float personal resentments and recriminations which may identify them. Instead I synthesise some of my reactions to the attendees' responses and moods. We were also faced with a question of method. How do you represent a non-place? How do you visualise a place no one claims to remember? We floated gloomy ideas, and then a man in his 50s suggested that we should represent precisely the gloomy, ghostly nature of the site. If we had to do it at all, someone suggested, we might as well attempt ghost photography. Not literally, the participant went on to explain, but out of pity for the poor souls who were left to rot in there for decades after the end of the war.

Next, we circulated the newspaper clippings and read them together. The room was suddenly quiet aside from the occasional tutting sound superimposed on the faint whisper of fingers on paper. Some had not read those news pieces and were full of anguish. A woman in her 30s admitted, "I just do not know what to believe". The words and images from the clippings acted as prompts for discussion but did not influence our conversation or our decisions to create the map; after all, those words were others' words; they did not belong to our group. While we could not ignore them, we could make sure these other discourses existed alongside out own. We then came to the other archival material we had available: documents from the regional archives and institutes. We decided to take the image of the only known victim whose photograph we had – Massimo Pollettini, a native of Belluno, a Blackshirt. It was decided to design a shrine around his image in the negative – a negative photo of a non-place, an anti-representation. Massimo's photograph bears witness to his being and not being there – his memory long purged. We took his image, solarised it and placed it alongside the cross on the plan of the sinkhole. Then we added the names of the known victims around the place: the ten Guardia Nazionale Repubblicana soldiers: Ruggero Ciruzzi, Riccardo Frare, Giovanni Brunello, Giovanni Birghin, Elio Romeo, Antonio Righetto, Antonio Brusadin, Giuseppe Boscarin, Sante Sperandio and Giovanni Rizzo. Then we added the only female victim, "Nella" Dal Bo De Pieri. We also had a couple of photographs of Nella. The former schoolteacher had found two Wiki-commons images of her that we could use without copyright infringement. We added those, also solarised, to the map draft.

Two participants (a married couple in their 70s) proposed that we add photographs of the speleological expedition that brought to light the corpses: images

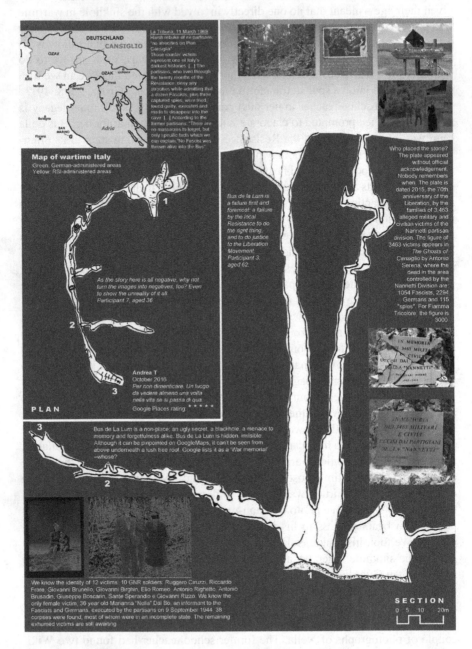

La Tribuna, 11 March 1969
Harsh rebuke of ex-partisans: 'no atrocities on Pian Cansiglio'
Those murder victims represent one of Italy's darkest histories. [...] The partisans, who lived through the twenty months of the Resistance, deny any atrocities while admitting that a dozen Fascists, plus three captured spies, were tried, found guilty, executed and made to disappear into the cave [...] According to the former partisans: "There are no massacres to forget, but only specific facts which we can explain. No Fascist was thrown alive into the Bus".

Map of wartime Italy
Green: German-administered areas
Yellow: RSI-administered areas

DEUTSCHLAND
CANSIGLIO
OZAV
OZAK
Adria
SAN MARINO

Bus de la Lum is a failure first and foremost: a failure by the local Resistance to do the right thing, and to do justice to the Liberation Movement. Participant 3, aged 62.

As the story here is all negative, why not turn the images into negatives, too? Even to show the unreality of it all. Participant 7, aged 36

Andrea T
October 2016
Per non dimenticare. Un luogo da vedere almeno una volta nella vita se si passa di qua.
Google Places rating: ★ ★ ★ ★ ★

PLAN

Bus de La Lum is a non-place; an ugly secret, a blackhole, a menace to memory and forgetfulness alike. Bus de La Lum is hidden, invisible. Although it can be pinpointed on GoogleMaps, it can't be seen from above underneath a lush tree roof. Google lists it as a 'War memorial' –whose?

Who placed the stone? The plate appeared without official acknowledgement. Nobody remembers when. The plate is dated 2015, the 70th anniversary of the Liberation, by the families of 3,463 alleged military and civilian victims of the Nannetti partisan division. The figure of 3463 victims appears in The Ghosts of Cansiglio by Antonio Serena, where the dead in the area controlled by the Nannetti Division are: 1054 Fascists, 2294 Germans and 115 "spies". For Fiamma Tricolore, the figure is 3000

IN MEMORIA DEI 3463 MILITI E CIVILI UCCISI DAI BRIGATA "NANNETTI"

IN MEMORIAM DEI 3463 MILITARI E CIVILI UCCISI DAI PARTIGIANE "B.G.A." "NANNETTI"

We know the identity of 12 victims: 10 GNR soldiers: Ruggero Ciruzzi, Riccardo Frare, Giovanni Brunello, Giovanni Birghin, Elio Romeo, Antonio Righetto, Antonio Brusadin, Giuseppe Boscarin, Sante Sperandio e Giovanni Rizzo. We know the only female victim, 36 year old Marianna "Nelia" Dal Bo, an informant to the Fascists and Germans, executed by the partisans on 9 September 1944. 38 corpses were found, most of whom were in an incomplete state. The remaining exhumed victims are still awaiting

SECTION
0 5 10   20m

Figure 5.10 Mapping a non-place: the visualisation of Bus de la Lum
Source: Copyright Sarah De Nardi

were available online. They had printed out two such images, and we gathered around to look at them: the photos dated from 1966 and were eerily serene, depicting the speleologists at the start of their exploration, perhaps unaware of the tragedy awaiting them below (Figure 5.10). We all agreed that these pictures should appear on our map, too.

The mood of a non-place: final considerations

Whenever we perceive someone or something to be wrong, out of place or aberrant, are we crafting and ratifying present and future prejudice? What if this prejudice redefines old meanings and past understandings of how the world works? How can we attempt to concretise and visualise stories about the war that automatically 'other' the outsiders? What if our audience were to change – say to include emotional outgroup members or 'neutral' individuals (say, foreigners)? How comfortable would we be with sharing feelings of schadenfreude about the enemy's demise or suffering?

And yet, this is what we are left with for the most part: melancholy, anguished and confused postmemories of a place which may or may not have been a mass grave. In attempting to process this very peculiar kind of mourning, we should interrogate the nature and extent of the materialities present and absent at the 'site' (Macdonald, 2009). The contrast between the human bones, transmigrated from the mass grave to a consecrated cemetery, and their "hauntological mystery is an uncanny absence" (Barbour, 2016: 95). Captivated by this tension, it is difficult to put into words the experience of being at Bus de La Lum. Simultaneously a natural and cultural locale, it is neither in human terms – it is an uncanny non-place, a black hole of memory. It is a focus of dissent, a vertigo of melancholia.

On reflection, this has proved the most difficult chapter to write, no, the hardest to even plan and execute, by far. There was a tension between the need and duty to tell the extraordinary Bus de La Lum story in a compelling way and my impossibility to get started. I found it hard to concentrate, impossible to put into words, to concretise, to 'render' my feelings towards this site on paper. The work with the community at Hotel San Marco may hopefully remain much more eloquent than the awkward prose of this chapter – which, I am aware, may come across as an exercise in awkwardness compared to the rest of the volume. And yet that event was worth reporting. That day we may have sat down and sketched our resentment, but we also tried to open up a space on the map for hope of reconciliation. An arrow or two, perhaps, that seek redemption and forgiveness, and try to lay the dead to rest, while attending to the living (see Figure 5.6). We decided to locate the photographs that witnessed the darkest themes (the victims) at the bottom of the sinkhole in our visualisation – almost as if the memory of the deeds had sunk with time and forgetfulness.

This chapter ends our foray into the historical consciousness of Italians who are still coming to terms with fratricidal violence during the war. It has not been an easy journey, but it has been useful, and hopefully fruitful to some. It has certainly led to some self-reflection – especially on my part. The visualisation experiments we have engaged in as Italians, as citizens and as descendants have sometimes served to open up our sense of identity beyond strictly textual and academic representations. People who do not like to read or do not have the time to do so have appreciated the notion that they could still get involved, learn and produce knowledge through other means. They liked the opportunity to work with others to construct a fairer, more open involvement with a past to improve their present. The visual experiments are not simply a metaphor for public history; to visualise something brings out things that are meaningful, even if they are hard to quantify or ephemeral. Storytelling in visual form can build on a politics of memory to create a reflexive future-making.

The visualisations of Bus de la Lum have brought the dead back in an acknowledgement of a lapse in humanity, for a brief moment in time. By attending to the side of our human nature that denies full, proper personhood and humanity to those who differ from us, we might perhaps overcome our tendencies to judge them as such. Italians should cultivate an awareness that the callousness deriving from infrahumanisation practices may blind them to the fact that others ('the Other side') also hoped, suffered and regretted things. In Italy, then, remembrance of the civil war could, as it has in Germany, lead the way to a long, painful but ultimately cathartic process of reconciliation (e.g. Tschuggnall and Welzer, 2002; Margalit, 2010). In the next section of the book, I turn my attention to the place-making mapping practices in the northeast of England. I engage with community memories and museum practitioners in the process of 're-imagining' and visualising vernacular memories of the 1950s in the region of the former coal mines.

Summary points

• In engaging with a painful place of anti-memory or divided memory, remote explorations are a viable alternative to site visits
• A cathartic reckoning with the histories and myriad legends tied to the sinkhole and alleged mass grave was the core aim to create this visualisation
• Stakeholders and community members volunteered their own thoughts and shared their own photographs and images of the site (in the public domain) to explore and channel how was it once was and how it looks and feels today
• The centrality of the victims' identities – and in particular the only female prisoner killed and buried in the sinkhole – was offset by an attention to the morphology and natural aspects of the site
• Attempting such a difficult visualisation can be a way to bring closure to a decades-long black hole in memory
• The courage in the face of possible social stigma of the participants must be recognised even if they remain anonymous

Notes

1 The Bus de la Lum is here: www.google.it/maps/place/Bus+de+la+Lum/@46.0591885,12.4079425,17z/data=!4m5!3m4!1s0x47790cab8320a7f9:0x78e4f7855e8f890e!8m2!3d46.0591848!4d12.4101312.

2 "Un luogo una foiba sicuramente da visitare [. . .] cavità naturale dove nel passato alcune di queste cavità vennero utilizzate in maniera impropria per far sparire delle persone".

3 "Un posto della storia e della sofferenza".

4 "Bella passeggiata in mezzo al bosco, qui si trova una voragine (inghiottitoio carsico) che fa un po paura!! Non si vede il fondo".

5 "Per non dimenticare. Un luogo da vedere almeno una volta nella vita se si passa di qua".

6 Stories from Beamish Museum's '1950s town'

Reanimating sense of place in the post-industrial northeast

This chapter takes us to northeast England, telling different stories shaped and shared by a vastly different set of communities. The chapter introduces us to a concept or idea not encountered in the Italian examples: nostalgia. In the English case, there is something to look back to with longing, as opposed to bafflement or anger. In framing this new element of the affectual assemblage of heritage, I ask whether nostalgia is more of a spontaneous affect or whether nostalgia occurs more frequently as an enabled performance enacted by heritage professionals and publics together in order to get the 'mood' right at a site or attraction. The case study in this chapter is about a special kind of place: a museum. The chapter traces the stages of the transposition of emplaced stories and material cultures from two real-world social contexts to museum reconstruction. In following the stories as they move (or not) from place to place, I frame these practices through the logics of heritage value co-production. Like heritage co-production, nostalgia may work as a connective tissue between heritage publics, practitioners, imaginaries and heritage objects through materiality and memory (Cashman, 2006). However, this observation positions us still ambiguously in relation to the spontaneous nature of nostalgia. "From books to computers, from mementoes to war memorials, material culture shoulders the larger responsibility of our personal and collective memory" (Buchli and Lucas, 2001: 80). The process of co-producing, of making and negotiating heritage values, may well rely on more than verbal clues and sensory experiences that exceed the discursive, narrative and representational. That said, how do museum professionals work with the absence of direct memories, and to what extent do the workings of nostalgia 'step in' to fill the gaps? How to coordinate the affectual performance of nostalgia-inducing heritage settings across participation by professionals and heritage publics? A trick museums use is to evoke, or even stage, feelings of nostalgia for a perceived shared past, or even to instil a longing for better, more innocent times now gone forever.

We may begin our exploration of nostalgia as an affectual manifestation of heritage by asking to what extent a forced or encouraged 'sensory' connection is pertinent or even applicable to replicas or reconstructions in living history museums. "The possibility offered by the museum is the world of imagination and, by extension, of potential with empathy by becoming other, if only momentarily" (Witcomb, 2010: 40). Witcomb elaborates on Ricoeur's (1996) crucial notion of narrative hospitality that, I argue elsewhere, may be the point of compromise between incompatible affectual understandings of history and a difficult past (De Nardi, 2016, 2019). As Gaynor Kavanagh observes, the museum is a place where irrationality marries the rational in a purposeful suspension of disbelief: the so-called museological dream space (2000). If memory is a mere copy, mementoes are copies of copies. But are these 'evocative objects' (after Turkle, 2007)? Elizabeth Edwards contends that mundane things (like photographs) can elicit deep emotional responses in those who engage with them: directly (nostalgia) or indirectly (imagined nostalgia, affectual outreach, which can have the same effect). Ordinary recalls merge with wider-spanning social understanding of time and history. Increasingly, "questions of history are at the heart of research, as are the politics of race, racism, equality, social justice and 'other' ways of experiencing the heritage landscape" (Tolia-Kelly, Waterton and Watson, 2016: 4). The expansion of the frameworks of the post-representational museum and heritage canons thus affects the wider social sphere with a promise of greater inclusivity.

We turn again to the senses as a paradigm for inclusivity and new, different ways of knowing: a museum is a place and as such can only be experienced through the medium of the senses and the sensing, affecting body social. Therefore museums are increasingly engaging with embodied and sensorial 'knowledge building' (Simon, 2008), after years of operating in pedagogical frameworks where "the senses were associated with the Other" (Witcomb, 2015b: 324). One exception to the historical reluctance to reengage the senses in learning may be open-air museums or living history museums, the history of which spans several decades across the world. Anthropologists of the senses Constance Classen and David Howes have written of open-air museums as places where "artifacts are displayed within a mock village, with typical houses, food, music, and inhabitant-guides" (2006: 218–219). Open-air museums seem to be the focus of the nostalgia industry thanks to ad-hoc engineered atmospheres and fictitious embodiments that enable experiential learning within their walls (or fences). The nostalgia industry is a manifestation of wider trends, a practice perhaps implicated in political and ideological processes which propose a return to the past as a solution to an unsatisfactory present and uncertain future (Dicks, 2015). Nostalgia may operate in isolation, singling out a community from the outside world or from certain elements within the community itself (Cashman, 2006). On another note, scholars and heritage practitioners have long asked what enables emotion in a museum setting: the agreement seems to be that a story well told, regardless of the significance of the objects or persons involved, is the most instructional as it sticks in the memory of visitors. This understanding follows Tilden's (1977) principles of interpretation: relating what visitors see in front of them with their own background experience (see Dicks, 2016). The unlocking of multiple perspectives not only conveys information, but also elicits emotional responses and awakens interest rather than serving a didactic purpose.

What can be learned, then, when the star attraction of a museum exhibit or story is nostalgia? Is nostalgia always and inevitably unruly and spontaneous, or can it be carefully staged and managed? Using the case study of the forthcoming 1950s town at Beamish Museum in northeast England, curated by the museum for its 1950s extension, I explore some of the affectual performances and imaginaries feeding into and seeping out of imagined communities of a 'nostalgic' past. The art of nostalgia, or rather its web of felt and staged affects in the museum space, serves as a paradigm for experiencing the past across multiple filters and expectations (Witcomb, 2015b). Is it possible to feel nostalgia for something you have never directly experienced or never really 'had'? This is important in thinking of the museum encounter at Beamish, styled to operate and appeal as "The Living Museum of the North". Who feels this fascination? Does the dream space trigger nostalgia (*sensu* Kavanagh, 2000) that unites a whole 'North' as an imagined community susceptible to the lure of a shared northern-ness, of a shared past materiality?

Bella Dicks (2016), reworking the Bordieuan idea of Habitus (1977), has made a convincing case for this concept to explore how visitors to 'everyday', social history or living history museums are themselves involved in making judgments and building connections with the display, not solely by appreciating their aesthetics, but also by becoming attuned to the social identities and experiential realm on display. Dicks argues that visitors position themselves in the assemblage forged by the display and the stories and materialities that form it, responding to it through their own 'autobiographical materiality' (De Nardi, 2016: 32–33). Visitors become part of the entanglement of identities, time and space in an active articulation of space that does not correspond to a passive construct of the habitus (after Skeggs, 2004). I find this perspective intriguing, and it is relevant to my own interpretation of the Beamish case studies that follow. In the community fieldwork leading up to the new Beamish exhibition, we too became variously entangled with the deeper-than-aesthetic qualities of places and things that make up the materiality and imaginary of the English northeast. In fact, only seldom have the objects' aesthetic been called into question.

Broadly, the dynamics of present and future entanglements with the memory and materiality of the 1950s in the northeast depend on many factors. The main museum tactic is to facilitate visitors' entanglements with their own habitus, enacting performances (Bagnall, 2003; Crouch, 2010), acting on their own post-memory, in order to enable meaningful encounters with the affects of an era still within living memory (Suleiman, 2008). Beamish Museum is adopting a similar strategy even to its more chronologically remote displays. How? The answer may be that the physical layout of Beamish facilitates rich, physical and emotional attunements for many of their visitors through a variety of sensory encounters. Such processes of "comparison and recognition, involving objects both present in the exhibition and others remembered and imagined by visitors, are as much sensory, embodied and imaginative as they are cognitive" (Dudley, 2010: 100). Thus, affect becomes part of the story told within the confines of Beamish and beyond: visitors carry their experience with them and talk about their visit long after they have left the compound, as it were, and get back to real life. In a few of the interviews I have carried out with Beamish visitors, it transpires that there is a fascination for the place – one research participant commented that "the museum tricks you, and you let yourself be tricked by it" – meaning, she went on, that you enjoy pretending that you have really stepped back in time (see Mulcahy, 2017 for a discussion of museums as liminal spaces of learning).

For Helen Graham, experiencing museums differently means to create museums differently (2016). This may entail the need to draw upon the stories told by real people in the real world, as opposed to the abstract knowledge building in the lofty ivory tower of the traditional top-down museum model. This methodology relies on the ethnographic encounter as the very foundation upon which to build knowledge – collaboratively (see Rose, 1997; Pink, 2009). I argue here that this endeavour is fundamental for the conceptualisation of the Grand Electric and the Airey Houses, evocative spaces for which multiple voices and perspectives exist. To frame the workings of nostalgia as a curated affect at the museum, I turn to Ricoeur's (1996) notion of the ability to "reach out" to others (persons, things and places) via our stories using "imagination and sympathy", for imagination and sympathy are powerful affects dictating our experience in every aspect of everyday lives. There are many cases where the narrative *trumps* the object: that is the case, for instance, in living history museums. After all, museums are the places where professionals tell stories – they do not teach (history, science etc.): "that's what schools do' (Blyth, 2015).

I believe it is very important to highlight alternative modes of learning at the museum which support alternative politics of participation, learning and knowledge construction – see Onciul, Stefano and Hawke (2017) for an inspiring collection that spans the spectrum of heritage participation techniques across various contexts. Other issues to bear in mind when imagining perception and learning at the museum include the following points:

- No one is born knowing how to use a museum
- Objects can tell a powerful story
- Objects can inspire
- Objects not only remind us of past practices and situations, but also act as 'portable places', transporting the self back to distant places and times
  The idea of the portable place (Bell, 1997: 821) offers a challenge for the design and making of memory maps 'representing' buildings due to be relocated to Beamish Museum – or any living history museum, for that matter. But what is portable? Affects, identities and memories, as well as bricks and other physical vessels for these connections?

A focus on the collaborative and bottom-up generation of knowledge is at the centre of the co-production project. Museum and heritage practitioners are responding to this sea change, if the collection by Onciul et al. (2017) is anything to go by. To this effect, co-creating knowledge is a dialogue that requires going beyond "message" towards "interactivity" (Barry, 2014). "A museum exhibition may 'act like a language' that 'prohibits or encourages different psychological processes' generated in accordance with the predispositions of visitors' habitus" (Dicks, 2016: 55). Museum visitors should be actively enabled to enact "bodily interactions with objects and discover how their experience differs from that of others who once used and made the displayed objects". Marguerite Barry has written on the pedagogical mileage of museum exhibits that introduce a 'please touch' strategy for visitors while most galleries and exhibition spaces still enforce a no-touch policy. But while no touching can be a distancing mechanism, the politics of interactivity are way more complex than that. If no touching is a barrier to engagement, learning and understanding, an invitation to touch can seem daunting too. "If the potential for interactivity is there but visitors do not participate and only observe the linear aspects of presentation, is the exhibit still interactive? Some visitors prefer to watch while others interact, but are they still participants in the communication too?" (Barry, 2014: 226). For Susan Dudley, practices of comparison and recognition, involving objects "both present in the exhibition and others remembered and imagined by visitors", are as much "sensory and imaginative as they are cognitive" (Dudley, 2010: 100).

The learning of the senses is more than just an emphasis on interactivity, however. Experiential engagement and learning are closely linked with social understandings of where one is in the world. Museum encounters plug into our experience as visitors and socialised human beings, bringing us closer to (or further removed from) the objects and stories on display. In this context, Connerton's notion of habit memories (1989) may also come into play. The remembering of experiences and tastes from one's past or even one's daily routine shape all new experiences and encounters. At Beamish, the making and consumption of fish and chips prepared a certain way, for example, may be a matter of a "remembrance in the hands" (1989: 93) in the context of the northern working-class identity. In this sense, there is a positive and generative potential to nostalgia as an affect beyond retrospective and conservative longing for a past: there are potentialities for present and future dreams. Food consumption and enjoyment pose yet another non-representational inroad to understanding spontaneous heritage encounters (Seremetakis, 1993, 1994, 2017; Korsmeyer and Sutton, 2011). The sensory experience and heritage of cookery is one of the most enticing aspects in the overall strategy adopted by Beamish: good food makes sense – literally. The affects circulating through and in 'productive' nostalgia and imaginations may thrive at Beamish Museum despite the lack of 'canonical' visual appeal of much of its displays and settings thanks to more ephemeral enjoyments, such as that of good food and beer.

Beamish Museum and the regional imagination: the maps as the foregrounding of a storytelling neighbourhood

In this chapter, besides analysing the pedagogical-affectual narrative that is the result of the move of tangible things from village to museum, I ask what happens when we extrapolate a building from its wider social geography. In so doing,

I frame the experience of bringing to life the memory of communities beyond a single household (the Airey Houses in Kibblesworth) or building (the Grand Electric in Ryhope) in terms of "storytelling neighbourhood" after the theory by communication scientists Ball-Rokeach, Kim and Matei, who articulate it thus: as "the communication process through which people go from being occupants of a house to being residents of a neighbourhood" (2001: 394). According to this research, storytelling neighbourhoods are the "most agentic process in the construction of those precious bonds that gestate coorientation in the form of imagined community" (2001: 394). What if we were to build on this idea of a storytelling neighbourhood as facilitating co-orientation, the bringing together of a world? What entanglements of identity, place and social memory might we foreground in the process?

We must start from the beginning. In visiting Beamish Museum in Stanley, County Durham, residents and natives of the northeast of England have an opportunity to engage reflexively and imaginatively in their own heritage. If we think of Beamish as an arena for the relatively unconscious enactment of playful, open-ended and unstructured autoethnographies, then these visitors-residents are excavating (so to speak) the remnants of the lifestyles of the local working classes wherein many locate themselves and their families. In using the term 'working class', I am mindful here of Skeggs's (2004) critique of a Bourdieuan framing of the working class as "lack, beyond value, without value, resigned and adjusted to their conditions, unable to accrue value to themselves" (2004: 87). The assemblage of individuals and affects that make up the local communities of the northeast both resist and exceed the paradigm of the disenfranchised working class trapped in their habitus (Benyon, Hudson and Sadler, 1986); rather, local communities come across as actively driven by a desire to learn about their history while living out their present identities in complex ways, both nostalgic and future-facing (Atkinson, 2007).

An autoethnography in this case is not a passive resignation to tried and tested patterns of social being and enacting a past, but an active and dynamic way that locals, as participants, connect their autobiographical stories to wider cultural, political and social imaginings and understandings. The term 'visitor' seems alien when talking about members of the local heritage and affectual 'ingroup', too, so a better word might be looked for – 'co-researcher' seems more apt. But is such an autoethnography nostalgic? The logics of affect may be seen to trump the 'passive' drive to nostalgia in creating wider scopes of engagement (Gregory and Witcomb, 2007); thus, Fenster and Misgav posit that "expressions of the spatial memory of a place emphasise similarities, differences and disagreements between people. They do not only reveal antagonism but increase awareness with regard to social relations" (2014: 365). A building of knowledge about the past through hands-on, reflexive, autobiographical engagements may indicate that even within a shared memory and construction of 'the local' there will be discrepancies and different ways of relating to exhibits, places and ideas presented to the visitors/ingroup members (Waterton, 2015; McAtackney, 2018).

I now move on to the forthcoming 1950s 'town' at Beamish. As an 'object', a restaged setting, much of the town may seem banal and inconspicuous to some who come across it. The familiarity of 1950s aesthetics might not seem 'interesting' enough to get excited about it, but they may elicit nostalgia and invite visitors to remember fondly their own or their parents' materialities – or to dismiss the clumsy, boxy furniture and tacky ornaments with a shudder. We might then ask what audiences will make of the humble, mundane and rather unglamorous materiality in the museum setting. The circulation of affects in the museum in and among bodies, things and staged settings will create a unique atmosphere: a simulated nostalgia for some and a buzz of fond recognition for others: an experiential rather than a pedagogic encounter (Gregory and Witcomb, 2007). The imagination and the enchantment pervading every aspect and actor in the project, from researchers and curators to the publics, is the key to an unprecedented exercise in place-making.

I selected as case studies these two respective communities that were the most interested in taking part in mapping and visualisation experiments with the museum, namely with Geraldine Straker at Ryhope and Lisa Peacock at Kibblesworth. Although at the time of writing the 1950s town is but a project, an imaginary in action, a place-in-becoming, the existing displays and staged settings at Beamish already speak to the visitors with a peculiar agency of presence. "Emphasizing the manner in which things create people is part of a rhetorical strategy to rebalance the relationship between people and things, so that artefacts are not always seen as passive and people as active. This will complicate notions of agency but allow us to make more of the rich analytical possibilities that artifacts offer" (Gosden, 2005: 194).

In our case, what rich analytical possibilities will 1950s kitchenalia, a bowling green and a few carefully placed lamps offer? Here we should look beyond the intrinsic value or putative representativeness of an object to 'connect' with its affective agency. To what extent can the mundane objects displayed at Beamish attain the same level of 'spark' or connection? So, for example, might someone say of a frankly ugly ornament from the 1950s, "Thanks for saving this (. . .). Someone will want it and cherish the fact you saved it" (Edwards, 2010: 31)? Communities "find it hard to surrender the past. Even while apparently engaging in planning, members may recall another era and imagine that remembering it will keep it alive or resurrect it" (Baum, 1999: 10, quoted in Fenster and Misgav, 2014). Affect becomes part of the construction of nostalgia operating within the walls on the museum. But does it feel familiar? Is it a genuine longing, and whose? More importantly, does reconstructing and repackaging everyday vernacular buildings as heritage sites work to encourage participation?

In the next sections I turn to the experiments with community memory mapping which, I argue, challenge the retroactive qualities of nostalgia to create vibrant present and future imaginings and plans. Co-curated memory mapping may bring together the imaginative, the creative and the unspoken in a story that 'illustrates' passions, dreams and convictions. And yes, a healthy dollop of nostalgia and make-believe too. Community-centred mapping, growing out of emplaced stories and affects, may channel the otherwise elusive imaginaries and the ephemeral (or haunting) mnemonic spaces encountered in fieldwork, which include spoken or unspoken longing for past identities and past practices (see also MacKian, 2004). Are these maps a kind of 'productive nostalgia' (see also Cairns and Birchall, 2013)? Co-producing maps of places that have existed, are now in-between and will surface once more at the museum allows us to engage with social movement mythologies, surfaced emotions and current concerns; ultimately, I believe the visualisations started conversations which will continue on the grounds of the museum.

The workings of heritage co-production we have glimpsed in the previous chapters have perhaps served to illustrate some of the possibilities offered by an affectual and imaginative engagement with the 'past' as perceived in the present. Mapping can tether these possibilities and actualise them in an accessible medium. Mapping heritage values and participation has many advantages: for one, through my work with the Engagement and Participation team at Beamish I was able to access and share stories and memories with local communities in the villages of Ryhope and Kibblesworth, but also to find strategies to visualise these memories, make them tangible once more (Figure 6.1). This will be more than an experiment with nostalgia: it will be about the reproduction and sharing of heritage meanings and memories created and shared through community and family at a local and regional level. In the latest map of the village of Ryhope (as published here) with its centre piece, the Grand Electric cinema (Figure 6.2), we can see the projected hopes, memories and dreams of a community that is no longer there, has

Figure 6.1 The Grand Electric before it was dismantled and taken to Beamish

Source: Copyright Beamish Museum (with permission)

Figure 6.2 The Grand Electric's interior: a stained-glass window

Source: Copyright Beamish Museum (with permission)

dispersed or has at any rate lost its economic and social core – a pit village without a pit, Ryhope is now a neighbourhood in south Sunderland.

The Grand Electric: a cinema, reimagined

When the Grand Electric cinema is dismantled and reconstructed at Beamish in the next two years, the affects and associations the building as a social core had to the community of Ryhope risk being lost. That is why I collaborated with the Engagement and Participation Team at Beamish to capture what life was like in the 1950s (see Chapter 7). The Grand Electric in Ryhope village, built in 1912, contained 900 seats in its heyday and operated as a lively and popular pit village picture-house. In 1956, the Grand implemented new technology CinemaScope: out went the square screen, in came the rectangular one. People in the northeast did not often go to the big, glamorous but pricey chain cinemas – they rather enjoyed outings to their local pit village picture-house. The Grand is remembered locally as an upmarket everyday picture-house, good for regular trips: our co-researchers recalled it as cosy, clean but unpretentious. "Meticulous", asserted former projectionist Bill.

I am interested in how the Grand Electric is remembered and how much of the popular imagination about the place will influence the remaking of it on the Beamish Museum's grounds. The reconstructed Grand is likely to resonate with the local community as most residents knew people working in the cinema. At Beamish, the plan and logistics of the reconstructed Grand have been tweaked, and the story the museum wants to get across may understandably differ from the memories of some of the former staff and patrons of the cinema. For instance, the auditorium and projection rooms constituted different worlds: the former projectionists and the community of cinema-goers of Ryhope remember them as being separate and self-contained spaces with their own independent operational economies and embodiments. The usherettes acted as intermediaries between auditorium and projection room, necessary to the economy of the cinema as the projectionists had no idea of what the picture sounded like. However, when the Grand Electric is reconstructed at Beamish Museum, the entrance to the projectionist's room (which was originally via a ladder) will be made easier, for better and safer access, through a door. This seems inevitable, but it will create linkages where there were none and channel communications and interactions that did not take place.

At our community meetings, which were held at the community hall in Ryhope and at Beamish Museum, the idea of a switch to digital film at the reconstructed Grand Electric did not resonate with the former projectionists, Eddie and Bill. "I want to get that feeling, of seeing reels in the analogue projectors", the two said unanimously. I thought about this: feels for reels. The museum intends to install at least one but ideally two 35 mm projectors. At the reconstructed Grand Electric, Beamish will need to work out the timing of the switch between projectors if they want to get that authentic feel. It took me some time to process this reaction to the museum's plan, and it has stoked my curiosity. The insider's view is always, somehow, going to differ from the outsider's perceptions – even with the best intentions on both parts.

While I thought about how to represent this divergence through our joint visualisations, my experienced Beamish colleagues tried to soothe some ruffled feathers by turning a perceived 'negative' into a positive and even a point of pride. Thus, the Beamish staff were keen to point out the potential 'boost' that the CinemaScope narrative will bring to their beloved former pit village picture-house. Beamish are keen to highlight the fact that CinemaScope was new at the time and that the excitement of the revolutionary innovation should be part of the story. Although the rectangular CinemaScope screen will be used, former projectionist Jack further commented that the square screen would be more in keeping with the classic look of the Grand Electric: "The square screen is more like the old times". This emphasis on the place as lived, not simply remembered, struck me as fundamentally important. This intuition led to related thoughts, which I sounded out with the community participants. The co-researchers agreed that the overall materiality of the place will feel different as well (Figures 6.3 and 6.4). The transformation from cinema to bingo hall in the 1970s meant that the interior was gutted and converted to an open-plan bingo layout.

The conversion entailed the removal and disposal of seats from the auditorium to make room for bingo tables and chairs. The gallery seats in the Grand Electric were still in situ, but the seats in the auditorium were gone, and seats donated by the former Palladium picture-house in Durham will end up filling the gaps in the

Figure 6.3 3D glasses from the 1950s found in the Grand Electric building during takeover

Source: Copyright Beamish Museum (with permission)

Figure 6.4 A film programme found by Beamish staff during preparations

Source: Copyright Beamish Museum (with permission)

reconstructed auditorium at the Beamish Museum. The local community stressed that different seating plans and configurations were important to the memory of the cinema, as they stirred up a sense of longing for childhood, teenage identity performances and social interactions: objects such as the seats may have emplaced an affectual economy of going to the pictures (see also Turkle, 2007). This gap between remembered lived experience and relocated sense of place stimulates the curiosity of my ethnographer self. Hybridity may compromise, if not confound, the habitus of encounter (Dicks, 2016). Local people or those who worked at the picture-house and bingo hall may experience disorientation in visiting the reconstructed building as an imbalance in their own habitus of the place. But this reimagining of a social space also intrigues the heritage professional: what place will the recreated Grand Electric be (Figure 6.5)? Will it be a heritage asset? The authenticity of the memories of the Grand Electric will have to be mediated for the modern world: the screen will not be nicotine-stained like in the original. The smoky atmosphere ("like a cloud", quipped Eddie) will be lost. Some cinemas had an extractor fan, but not 'the Grand'. A woman remembered how her uncle had gotten lung cancer from the passive smoke lingering in the projectionist's room. Indeed, local ex-projectionist Jack admitted that on occasion the picture became blurred due to smoke: "It was murder to try and focus", he said. This made us think: while I feel this sanitisation of the past is essential to the story as children nowadays live in a 'smoke-free world' and have no notion of how the old times felt – smoke choked and smelly – young people still deserve to learn about the sensory experience of the former mining communities, including the evocative smokiness of the 1950s.

Legend to the Ryhope Map

1 **The Grand Electric:** Cinema and bingo hall. Built 1912.
2 **Burdon (sp) Terrace and Tunstall Street** are vanished terraces of colliery houses. Ryhope colliery opened in 1857, and these houses would have been built near the mine for the pitmen and their families; earlier, Ryhope had been a small village. The spelling of 'Burdon' is variable, and likewise the local geography has changed. There was a *Double Burdon Street*, which was to the north, and parallel to Ryhope Street, the B1286. *Single Tunstall Street* was between them. Former residents include Doris (Burdon Terrace), Jenny, b. 1959 (Burdon Street). Brick Row (still there), which led to Ryhope Colliery, ran across the west end. On the modern map there are now Western Hill and Shaftesbury Avenue.
3 **Jock's Ice Cream Shop:** This was on Ryhope Street near the Western Hill junction. People loved congregating there on a Sunday after going to the Grange Town Regent Cinema. The Grand Electric was built on church land, so it did not open on a Sunday.
4 **Store Fields:** A playground the children used to go to in the 1950s. Every year Clarks Fair used to come with a climbing frame and all sort of rides and attractions like the teacup lids.

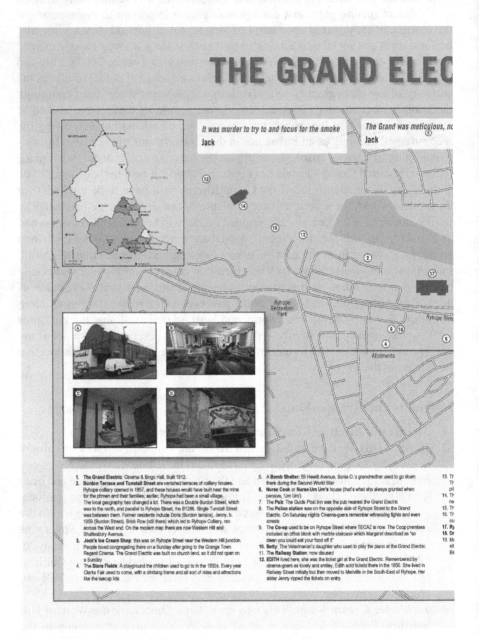

Figure 6.5 A community map of Ryhope

Source: Copyright Sarah De Nardi

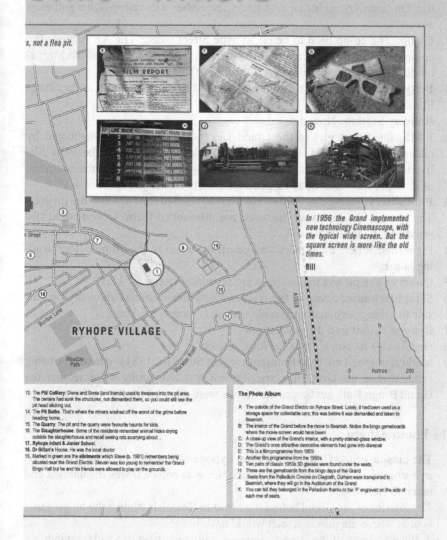

CTRIC – RYHOPE

s, not a flea pit.

FILM REPORT

In 1956 the Grand implemented new technology Cinemascope, with the typical wide screen. But the square screen is more like the old times.

Bill

RYHOPE VILLAGE

Meadow Park

0   metres   200

13. The **Pit/ Colliery**: Diane and Sonia (and friends) used to trespass into the pit area. The owners had sunk the structures, not dismantled them, so you could still see the pit head sticking out.
14. The **Pit Baths**. That's where the miners washed off the worst of the grime before heading home...
15. The **Quarry**: The pit and the quarry were favourite haunts for kids.
16. The **Slaughterhouse**. Some of the residents remember animal hides drying outside the slaughterhouse and recall seeing rats scurrying about...
17. **Ryhope Infant & Junior School**.
18. **Dr Gillan's House**. He was the local doctor
19. Marked in green are the **allotments** which Steve (b. 1981) remembers being situated near the Grand Electric. Steven was too young to remember the Grand Bingo Hall but he and his friends were allowed to play on the grounds.

The Photo Album

A: The outside of the Grand Electric on Ryhope Street. Lately, it had been used as a storage space for collectable cars; this was before it was dismantled and taken to Beamish.
B: The interior of the Grand before the move to Beamish. Notice the bingo gameboards where the movie screen would have been!
C: A close-up view of the Grand's interior, with a pretty stained-glass window.
D: The Grand's once attractive decorative elements had gone into disrepair
E: This is a film programme from 1955!
F: Another film programme from the 1950s.
G: Two pairs of classic 1950s 3D glasses were found under the seats.
H: These are the gameboards from the bingo days of the Grand
J: Seats from the Palladium Cinema on Claypath, Durham were transported to Beamish, where they will go in the Auditorium of the Grand
K: You can tell they belonged in the Palladium thanks to the 'P' engraved on the side of each row of seats.

Figure 6.5 (Continued)

5   **Bomb Shelter:** 59 Hewitt Avenue. Sonia O.'s grandmother used to shelter down there during the Second World War.

6   **Nurse Cook** or **Nurse Um Um's** house (that's what the nurse always grunted when pensive, 'Um Um').

7   **Pub:** The Guide Post Inn was the pub nearest the Grand Electric.

8   **Police Station:** On the opposite side of Ryhope Street to the Grand Electric. On Saturday nights cinema-goers remember witnessing fights and even arrests.

9   **Co-op:** The co-op used to be on Ryhope Street where TECAZ is now. The co-op premises included an office block with marble staircase which Margaret described as "so clean you could eat your food off it".

10 **Betty:** The veterinarian's daughter who used to play the piano at the Grand Electric.

11 **Railway Station:** Now disused.

12 **Edith**, the **ticket girl** at the Grand Electric, lived here. Remembered by cinema-goers as lovely and smiley, Edith sold tickets there in the 1950. She lived in Railway Street initially but then moved to Mariville in the southeast of Ryhope. Her sister **Jenny** ripped the tickets on entry.

13 **Pit/Colliery:** Diane and Sonia (and friends) used to trespass into the pit area. The owners had sunk the structures, not dismantled them, so you could still see the pit head sticking out.

14 **Pit Baths:** Where the miners washed off the worst of the grime before heading home.

15 **Quarry:** The pit and the quarry were favourite haunts for kids.

16 **Slaughterhouse:** Some of the residents remember animal hides drying outside the slaughterhouse and recall seeing rats scurrying about.

17 **Ryhope Infant and Junior School**

18 **Dr Gillan's** house: He was the local doctor.

19 Marked in green are the **allotments** which Steve (b. 1981) remembers being situated near the Grand Electric. Steven was too young to remember the Grand Bingo Hall, but he and his friends were allowed to play on the grounds.

The Photo Album

a   The outside of the Grand Electric on Ryhope Street. Lately, it had been used as a storage space for collectable cars; this was before it was dismantled and taken to Beamish.

b   The interior of the Grand before the move to Beamish. Notice the bingo game boards where the movie screen would have been!

c   A close-up view of the Grand's interior, with a pretty stained-glass window.

d   The Grand's once attractive decorative elements had gone into disrepair.

e   This is a film programme from 1955!

f   Another film programme from the 1950s.

g   Two pairs of classic 1950s 3D glasses were found under the seats.

h   These are the game boards from the bingo days of the Grand.

i   Seats from the Palladium Cinema on Claypath, Durham, were transported to Beamish, where they will go in the auditorium of the Grand.

j   You can tell they belonged in the Palladium thanks to the 'P' engraved on the side of each row of seats.

Reimagining the Airey Houses

The Airey Houses, Kibblesworth, were two-storey semi-detached dwellings with a metal skeleton made from the recycled frames of military vehicles clad in slats of greyish-brown concrete. In the case of these buildings we can map divergent opinions and imaginaries in turn as insiders and outsiders engage with the presence/absence of the prefabricated homes (Figure 6.6). The Aireys have now been pulled down, but they remain as a mixed-feeling artefact in the local landscape beyond the memory of their physical form.

Even the logistical configuration of the Aireys is disputed. In the course of their oral history research, Beamish Museum found that most people associate the bungalow format with the notion of prefabricated dwelling – not the two-storey semi-detached homes acquired by the museum and that were once scattered in the wider area. Whatever their configuration, however, there is agreement in an outsider perspective that resented the buildings' presence: Gateshead and Team Valley newspapers and vox pops firmly hold the view that they were an eyesore and that people were glad to see the back of them. In a newspaper article from 2012 we read:

> Much less appealing superficially than the bungalow version, these two-storey semis [. . .] shared the bungalows' advantage of mod cons but were much disliked as a blot on the landscape. One of my great uncles started a speech to the Yorkshire branch of the Institution of Civil Engineers back in the 1950s; "*Ladies and gentlemen, except Sir Edwin Airey. . . .*" Family legend likes to believe that this was more because the builder was a vigorous Conservative, but there was certainly an element of anger at the sea of concrete.

In the oral history interviews, however, I met nothing but affection towards the Airey Houses from their former residents. The maps I sketched with the community, therefore, seek to reconcile the two sets of memories, but with a focus on the insiders' perceptions, as they were the ones who lived there in the first place. On 6 December 2015 held at the Kibblesworth village hall, our research team tried to figure out how we could best map out their village landmarks. We took a rambling walk through snowy Kibblesworth that day. As co-researchers walked me through the village, they positioned their memories and the 'absent' Airey homes. They pointed out other standing buildings, such as the boarded-up co-op (the general store). We lingered at the miners' club house. Here Colin, a non-Airey resident in his late 50s, had explained that "it was built by our local miners. They built the ground floor at first. The local miners then had to go around to the other clubs in other villages, and they had to win at dominoes etc. to get the

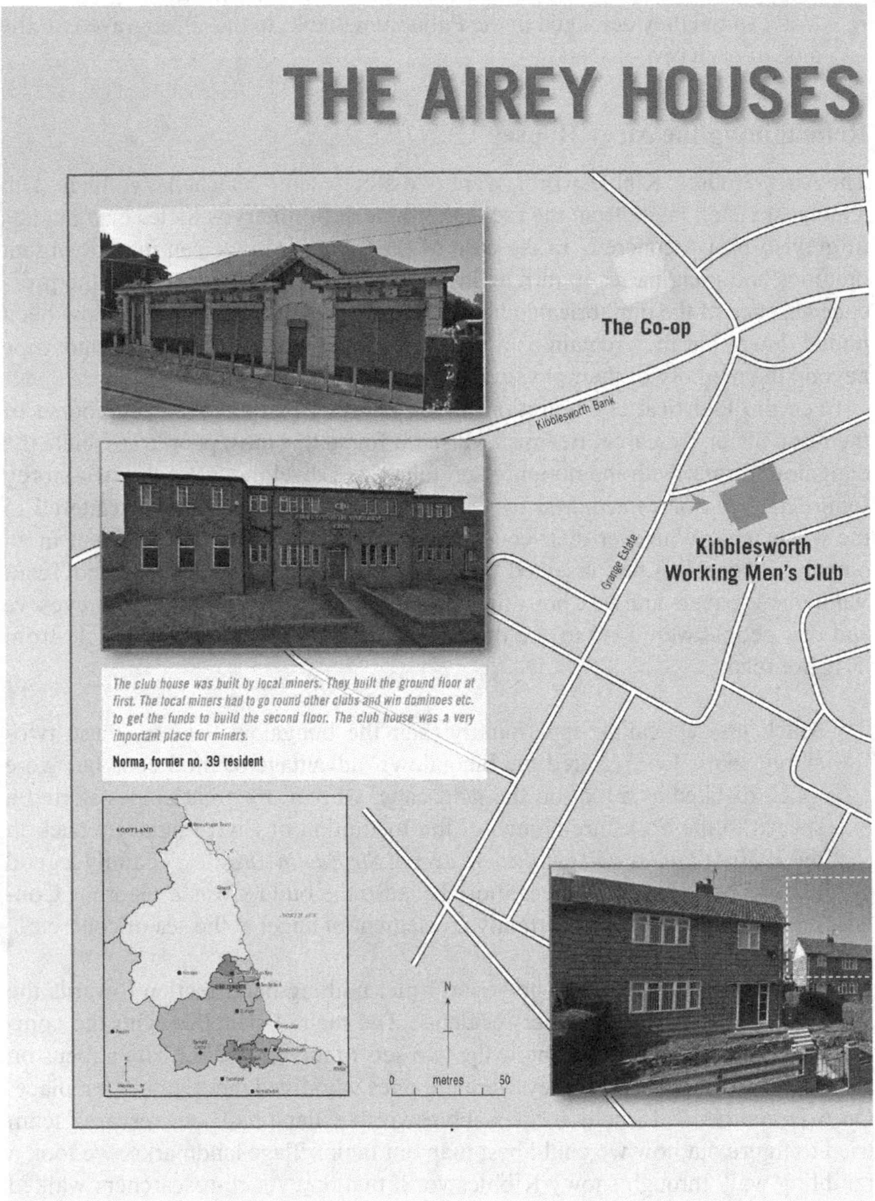
THE AIREY HOUSES
The Co-op
Kibblesworth Bank
Grange Estate
Kibblesworth
Working Men's Club
The club house was built by local miners. They built the ground floor at first. The local miners had to go round other clubs and win dominoes etc. to get the funds to build the second floor. The club house was a very important place for miners.
Norma, former no. 39 resident
SCOTLAND
N
0 metres 50

Figure 6.6 A Kibblesworth memory map

Source: Copyright Sarah De Nardi

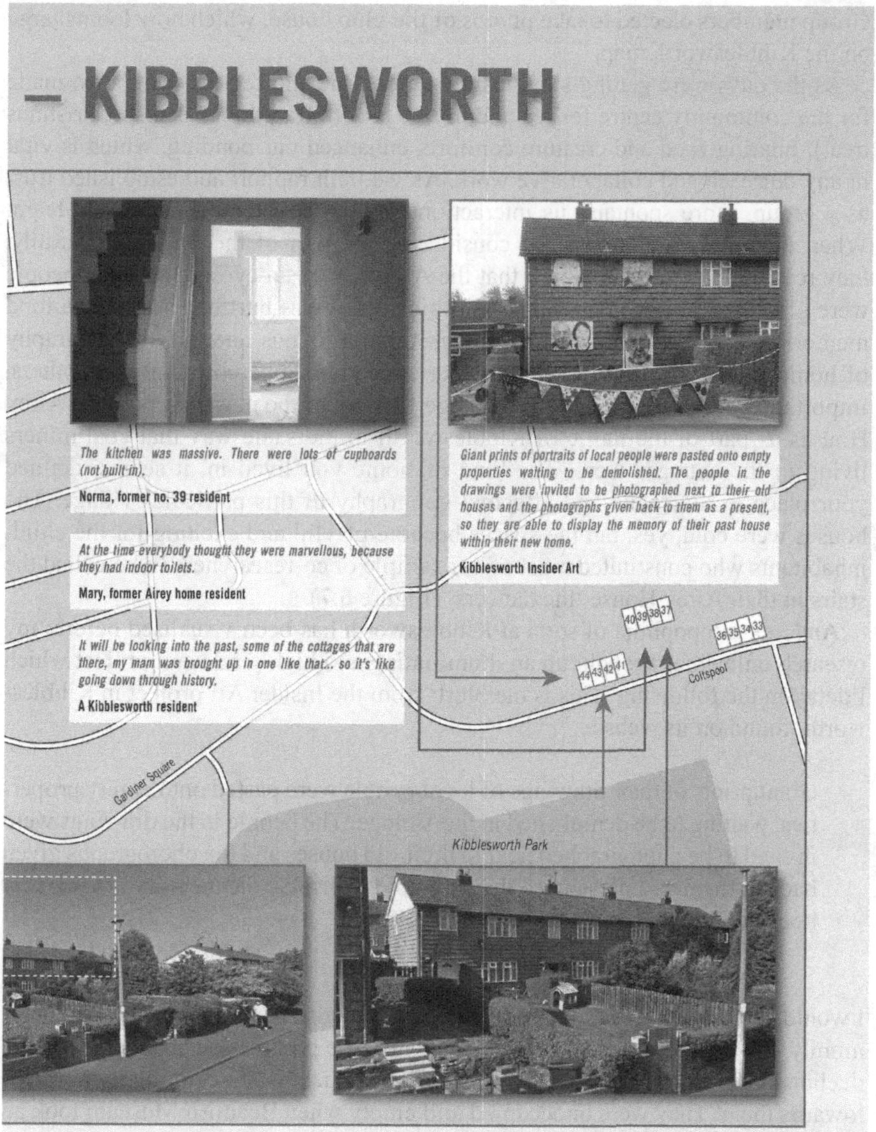
– KIBBLESWORTH
The kitchen was massive. There were lots of cupboards (not built in).
Norma, former no. 39 resident
At the time everybody thought they were marvellous, because they had indoor toilets.
Mary, former Airey home resident
It will be looking into the past, some of the cottages that are there, my mam was brought up in one like that... so it's like going down through history.
A Kibblesworth resident
Giant prints of portraits of local people were pasted onto empty properties waiting to be demolished. The people in the drawings were invited to be photographed next to their old houses and the photographs given back to them as a present, so they are able to display the memory of their past house within their new home.
Kibblesworth Insider Art
40 39 38 37
36 35 34 33
44 43 42 41
Coltspool
Gardiner Square
Kibblesworth Park

Figure 6.6 (Continued)

money to build the second floor."[1] The members of the research group felt this story was extremely important to 'get' the community spirit of Kibblesworth. Group members elected to take photos of the club house, which now looms large on the Kibblesworth map.

As the days were getting significantly colder and shorter, the team soon made for the community centre for tea and mince pies (a traditional British Christmas treat). Sharing food and creature comforts enhanced our bonding, which is vital in any outreach and collaborative work. As we built rapport and established trust as a group, more spontaneous interactions started developing within the team. When the co-researchers started considering me part of the 'ingroup', an ally, they revealed the public opinion that the Aireys were an eyesore and that people were glad to see them gone. Some felt this stigma was hurtful. Instead, the fond memories of the prefabricated dwellings trace a joyous and playful geography of home in stark contrast with the experienced/remembered ugliness of others. Importantly, outsiders never acknowledge the role of the newcomers to the Airey Houses as part of the fabric of Kibblesworth in the same way that coal miners living in the cottages were. The kind of home you lived in, it seems, defined your place in the social and affectual geography of this particular village. The houses were cold, yes, but the spaces became playful and exciting for the child-inhabitants who constituted most of our sample of co-researchers (who called the stairs in their Airey House 'the dancers' (Figure 6.7).

And yet, a topophilia of sorts at Kibblesworth has been visualised before: my research online has brought up an illuminating and highly relevant project which I detail in the following. This is the blurb from the Insider Art project in Kibblesworth, found on its website.[2]

> Giant prints of their drawings of local people were pasted onto empty properties, waiting to be demolished in the Village. The people in the drawings were invited to be photographed next to their old houses and the photographs given back to them as a present, so they are able to display the memory of their past house within their new home.
>
> (Rednile Projects, 2018)

I would call this an act of love for a building, but also a strong bond among community, the village and the Aireys themselves. The progressive abandonment and decline of the building do not detract from the attachment a community nurtured towards them. They were unoccupied and empty when Beamish Museum took an interest in them, and their apparent life cycle had come to an end, but the Aireys were not unreceptive empty shells. Life went on in and around them still. Navaro-Yashin (2012) has written on the pull and allure of the phantomic of places after they have been abandoned – the afterlife or new life of things whose users or owners left. DeSilvey (2006) has also reflected on the power and meaning of decaying things: for these scholars, the traces of things that once 'were' is as potent and as creative as preserved, functional objects. So, decay is 'storied' in its very occurrence, as a thing, a process, a life stage of an object. Usually, such

Figure 6.7 The dancers – inside an Airey House

Source: Copyright Beamish Museum (with permission)

life cycles thrive and bloom even after places or objects evolve into disrepair or new life within reach of persons or non-human animals with which they interact.

But what if the things in question languish in storage or are absent, disconnected, uprooted from organic and inorganic interactions? The maps I made with the community evoke local materialities and their memories of a domesticity long gone. When the Aireys come out of the darkness of storage and see the light of day once more at the museum, the prefabricated semis will be able to tell the story of a community and a village through the maps made with the community. They will be there as a tangible presence to an absence, the vibrant mining community they were once a part of despite mixed feelings about their aesthetics (see Dicks, 2015).

An elusive future: Esther G.'s Sunderland home

A tethering of nostalgia

Now I return to the idea of nostalgia. The next logical question might ask for whom nostalgia emerges. As a largely singular and intimate feeling, nostalgia may be reliant on specific parameters that trigger a 'fuzzy' feeling of longing for a past: the time and place have to be 'right'. A site or a house museum is rather a "three-dimensional archive; a phenomenological journey through the spatial layerings of another's personal archaeology" (Hancock, 2010: 114). Will Esther's Sunderland Council semi-detached dwelling evoke nostalgia in most visitors? It is hard to say as the building's journey of reconstruction is just beginning. The rebuilding of Esther's home, as a project, encompasses a whole spectrum of affective responsibility, representation, credibility, uncanniness and love. But will it channel nostalgia? Chosen by the people in a public competition, this house will contain no elements that are at odds with its inhabitants, but it will be a replica nonetheless. It is going to be a rare breed of house ecology: neither house museum nor tribute, but rather a doppelgänger in the staged museum setting of the 1950s town. Wedged between the police houses and the allotments, the replica-house will have to compete for visitor attention with the cinema, the fried fish shop and the bowling green: more public spaces, less imbued with emotional resonances and much less redolent of domesticity.

In filtering our impressions of the 'Sunderland Semi', we are left with more questions than answers. Will this be a domestic space in which strangers would take no interest or even feel like intruders? Is such a home setting likely to resonate more with women, with families, with older people? If objects are "multisensory embodiments of meanings" (Classen and Howes, 2006: 201), what embodiments will be encountered in the house-to-be? The anticipation of the creation of its benign doppelgänger in the museum is an affect in itself, a thing in becoming which has taken on a life of its own in the process. Despite protestations of authenticity or otherwise, we know that the house, and some of its contents, will become part of another place's cultural geography. The outside and inside of this particular house become part of a different affective and experiential configuration as part of the 1950s town. Esther claims to be delighted to become part of the history of the museum and that people will 'see how they lived', but will they really? What expectations does Esther's family nurture?

With its past inhabitants unable to live in the recreated home the museum will have to work with other affectual triggers to make visitors take an interest. Things will have to take centre stage. They will have to master and convey nostalgia for cosier homes or simpler times. "By emphasizing the movement across bodies, trans-corporeality reveals the interchanges and interconnections between human corporeality and the more-than-human" (Alaimo, 2008: 238). A handful of personal belongings will become part of the house's story on the stage set at Beamish. Esther's family has agreed to donate a selection of objects from the 1950s to become part of the interior of the recreated family Council house in the museum. But will these things make themselves at home again? Will the house be a home? Here we may reflect on the characteristics of 'home' as an encounter, as a node of affects, perhaps even nostalgic, which may or may not be part of the experience of visiting someone else's home. The house museum (for a rich literature on this see Alcock 2010, Kavanagh, 2000) becomes an arena in which we reach out to others' stories via our own stories (Ricoeur, 1996). The connection is established on multiple levels via our own personal stories, our previous experiences of place and our assumption that someone, once, lived in that place. The house museum is, or was, someone else's prison, shelter, haven or corner of peace and quiet. It could have been a locus of violability, of misery and the epitome of personal destruction. And yet, good or bad, the house was a place close to someone's imagination, part of the local chorography to someone. That is why a house that has been lived in communicates with us. Lucy Lippard has called this 'pull' the lure of the local. For Lippard, Esther's recreated semi-detached dwelling will not be 'placeless', as she defines placelessness as what is "unknown, unseen or ignored" (2000: 9). In what follows I examine the affects which will be potentially evoked by the 'Sunderland Semi' at Beamish in light of two of these anti-criteria: the unknown and the unseen.

The unknown

Nuala Hancock (2010) identifies a house museum as a journey though someone else's personal archaeology. Here I take this statement to mean that to explore Esther's home is to experience, albeit vicariously, the accretion, conglomeration and coming apart of another person's possessions. The geography of the house speaks to us in its uncanny familiarity. Even if we have never set foot in this particular home, it resonates with us. Why? In light of its atmosphere, its affects, its smells and sounds and textures: the soft carpet underfoot, the familiar sight of tea towels. When we tread onto a creaking floorboard, the familiar sound will bring us back to a number of old houses we have ever set foot into. That is the focal difference between Esther's family home and its future doppelganger in the museum.

The recreated domestic husk of Esther's family home at Beamish will be lacking that very uncanny, ineffable quality that makes a house a home – its lived in, echoing rooms and stuffy corners; the creaking of well-treaded floorboards; the thin fragrance of dinner cooking on the stovetop; the warmth of bodies and the excitable chatter of children. The smell of the horse manure at the front of the house after the cart has come and gone, and Esther's much joked about role in shovelling up the droppings – these details will be lost, part of the story lovingly told during reminiscence sessions, shared among the family, cherished as yet another layer or texture of the affectual history of the house and its occupants.

But what if the things are absent, ignored, far away? Do echoes of lives and livelihoods in a building like the Grand Electric survive in an artificial, recreated setting? Does the cupboard with a door that always sticks from back home on Coltspool Lane make it 'whole' to Beamish Museum? Might the sour smell of the compost heap in the garden reach our nostrils across time and space? The sights and sounds of a life well lived: where are they? Can they be moved? Are they part of the fixtures, fittings, chattel adorning a house and making it a home?

The unseen

What about the things that tell a story and communicate affects to other people but will remain at home, in Sunderland, and *become unseen* or unexperienced in the museum version? The human and more-than-human dimensions of a life lived in full in a house (the gardening, the shopping trips, the horse shit) become part of what Navaro-Yashin (2009) has called a ghostly economy, a phantomic make-believe space. The make-believe space at Beamish encompassed recreated husks, supposed to contain the stories and affects within the museum compound, relying on the materiality of the recreated setting to tell a story which will interest and stimulate visitors. That once familiar Sunderland street will not be there anymore outside the house's doppelgänger at Beamish; another ad-hoc ecology of street and neighbours will surround the projection of this family unit. And yet we will, nonetheless, feel the pinch and echo of the familiar in no uncertain terms. The chatter of tea cups and television sets will be long gone, but we can picture ourselves treading its oddly unfamiliar floors. We can imagine ourselves walking up the front path, carrying good or bad news, surely?

In terms of unknown and unseen, we are not only dealing with the imaginaries of what Esther's home will be. With the Grand Electric in Ryhope, we are also feeling our way in the dark. The museum just does not know the full story, although the Grand is still (at the time of writing) standing in its corner of Ryhope. This building was an odd one, to say the least. As Geraldine Straker explained to me (see following chapter), not all phases in the life and use of the building are well understood or even fully known to the experts, owners or local residents of Ryhope. Some of the phases of use or modification from cinema to bingo hall to garage are still unexplored. For instance, no one knows with any degree of certainty when the current 'new' façade of the Grand went up or when the access to the projection room was modified.

Imagination and forensic exploration during the dismantling of the structure have to fill in the gaps in knowledge. At that point in time Beamish did not know the full history of the building, nor why the 'relics' (3D glasses, movie programmes) had not been disposed of by the building's owners. Archaeologically, and socially, there was certainly a destructive/generative aspect to the dismantling of the cinema and its move to Beamish. The act of dismantling would generate knowledge, add something to the derelict structure and lend it a new lease on life, but also disrupt a community that had developed around it. The act of systematic demolition would destroy a place *in place*, yes, but also lead to the discovery (one hopes) of its secret during the act of peeling back, dismantling, looking behind and underneath what is visible now. The relocation to the museum would peel back the layers of the Grand's existence: a learning-through-unmaking.

My imagination was tickled by the idea that the community, the museum and I were making maps about places in between, hidden and yet present in the imagination and in memory, using elements that came from all of these dimensions – place, memory and the imagined. Mixed-up memories of the cinema populated the reminiscence sessions and workshops with the community. Museum staff acknowledged with an extent of fascination that, even when exploring the 1950s cinema specifically, the memories they collected were not entirely of that era. The stories they got often encompassed periods and even people who knew and used the Grand afterwards, in the 1960s, 1970s, and 1980s, when the venue was a bingo hall. In terms of the map-making, narrating and visualising the memory *and* the imagined at Ryhope was a collaborative endeavour of discovery, trial and error. The maps, and the archaeological uprooting by the museum of Grand Electric and the Airey Houses, is an act of disclosing of the unseen, the forgotten and the absent/present sense of place of two distinct communities.

Conclusions to a colourful journey in the present-past

Back to the Red House – Esther's home. We could not know then how Esther's story would unfold in the reimagined rooms of her fictitious home, and I am writing in an uncanny non-time, a point in time when the house had not begun to take shape but of course future readers could physically go and enter it. Nor do we know how Esther felt in each room, opening the curtains on a given morning, or priming their *thingly* contents for a special occasion, or decking them with Christmas lights when the nights became bitterly cold. The recreated house at Beamish Museum will not be ignored – it will be a thing – but how do we conceptualise the things about its history that we ignore? Does that ignorance even matter? Once it resurfaces at Beamish, how much of Esther will we encounter in the make-believe space of the replicated Sunderland Semi? Perhaps the in-betweenness of the state of that house at the time of writing troubles me. It may be that I have to fill the space and time with chatter in order to ignore the non-existence (at the time of writing) of the doppelgänger at Beamish. Is the phantomic state of the make-believe homeliness of the Esther's family unit haunting me? In deciding to write an elegy for a place that does not exist yet except as a blueprint and in the imagination of Esther's family and the Remaking Beamish team at the museum, I perhaps do it a disservice. Just because it is not a physical place yet, one composed of bricks and mortar, does not mean the in-between Sunderland Semi does not occupy a space in the emotional stakes of the project as a home of sorts. We can trust the imagination and memory to populate this home and to fill in and give life to other unspecified spaces to come.

As in much social history–facing heritage work, nostalgia may well become a veneer in the representations that community mapping generates at Beamish

Museum and beyond, but not in any stagnant way. Unlike in the Italian case studies, the past is not painful here: it is a comfortable feeling of going home. A home that is now going to be recreated in a fictional space of collective experience of the 1950s, there to be felt and explored by young and elderly alike; a dream space soaked with expectations and local colour. In this dream space, imagination will be key. It will connect the dots, fill in the gaps in memory for those too young to remember or those, like me, whose cultural background and materiality is utterly different.

The next chapter, a section of which is co-authored with Geraldine Straker of Beamish Museum, illustrates how we conceptualised and populated two 'deep maps' with community co-researchers. We did not simply attempt to capture oral histories; rather, we planned to reconcile and visualise different sets of memories and imaginaries; among these, affectual links to place in the original villages, shared imaginaries negotiated during the reminiscence and oral history sessions and mapping workshops and the new life of materialities *over there*, already (not yet) at Beamish. For us as researchers, all of those elements were embroiled in a learning experience made of competing yet complementary understandings of what these places were, what they are now and what they can become. The colourful in-betweenness of the Grand Electric and the melancholy incompleteness of the Airey Houses and Esther's Sunderland Semi appear to us as a blueprint for change, for a future shaped by the imaginaries of many. Like the other maps we have encountered in the book, the focus here is on the insiders' perceptions, as they were the ones who lived or still reside in Ryhope and Kibblesworth in the first place. These are their maps, not mine or the museum's. As for the Grand Electric cinema, the projectionists' ambivalent feelings towards film and screen formats relay a core message which informs the memory maps we devised as a group: that they ultimately own the stories. The Grand Electric's story is a variegated and colourful mélange of people, materialities and stories from the community, which we tried to convey with the multi-layered map – a collage of recollections and sensory worlds that may become lost.

The collective remembrance of the coal mining industry situates stories and values through collective knowledge and exchange. The maps I initiated with the local communities bring up more than data and historical context; the visualisations disclose the emotive undercurrents of sites which have been or will be destroyed, uprooted and moved to Beamish Museum. The communities I have worked with so far have thoroughly enjoyed the idea of putting their thoughts and memories on a visible, accessible medium which can be copied, circulated and emulated by other communities telling their own stories. There is scope for the maps and visualisations to become a method schools use with young pupils in the creation of 'sense of place' and as a classroom methodology in social studies and the humanities.

The maps devised with Beamish Museum and the local communities will hopefully remain on site to tell their tale, in some form or other. They will be likely modified, expanded, reimagined: that is their fate, and that is what they must do. The maps in their current and future form will convey meaning as an homage and testimony to the various locales where many affects took place and intermingled – their humble representations truer to life than any oral history transcript or museum label could be and much more vibrant and socially generative than nostalgia. I wonder if the emphasis on the maps at Beamish Museum, where they may be displayed as large billboards or displayed at times so they can be part of an ongoing conversation, will ease the focus on 'message' and instead perhaps shift visitors' attention more towards the lived experience, the *living* and future-facing memories of northeast communities. The imagination (an element within the creation of nostalgia) may be a "collective practice that operates in ways similar to those suggested for collective memory" (Pink, 2009: 45). Nostalgia may also be a force for retroactive exclusion process rather than inclusion and reconciliation, triggered by a desire to return to an "unsullied past" (Bonnett, 2010). Either way, imagined or longed-for ways of life come across as dynamic, not still. They are an act of doing, of vindicating communities' special spatial places. In imagining the kind of community the museum's recreated 1950s will be, we may think of it as a path to togetherness.

Summary points

- There is more to the social memory of the English northeast than a hagiography of the coal pits
- The entertainment industry and the domestic everyday lives of the recent past are equally important
- Communities in the northeast reflect on their recent past by talking about their expectations for representation in the present and their future aspirations
- Inclusive and multivocal memories intersect place on the vernacular maps
- The two maps of Ryhope and Kibblesworth serve two different purposes: the former serves to bring to life the vibrant community around the cinema which, in isolation, will re-emerge at Beamish Museum; the latter vindicates sense of place at a village which nurtures positive memories of buildings that outsiders view as ugly and unappealing

Notes

1 Colin, interview, 3 March 2016.
2 The Insider Art @ Kibblesworth is an initiative by Rednile Projects 2018. All rights reserved. Available at: www.rednile.org/public-realm/insider-art-kibblesworth/ [Accessed 21 July 2019].

7 Experiencing the mapping method in the field

A dialogic interlude

Co-authored with Geraldine Straker, Beamish Museum

Introduction

Community is a way of learning. Community is also a communication process: it is a way of learning by doing, by being together, by sharing imaginaries, hopes, dreams and plans in a polysense (after Sather-Wagstaff, 2016). A community is a learning curve in the sharing of a past, too. A way to intersect these various ways of learning is to find a linkage among materiality, memory and the liminal space of heritage. Museums have the fundamental role to educate about the past while being grounded in the present, and this takes place in multiple concurrent ways (Simon, 2008). As liminal spaces of learning (Mulcahy, 2017), museums inhabit uncanny states of in-betweenness: in between history and memory, materiality and affect. As political institutions immersed in the social, museums often need to narrow down their representational tactics to one predominant narrative tack while maintaining an ethical and transparent political positioning and an awareness of multiplicity (Castañeda, 2008; Hall, 2005). The liminality of museums (Mulcahy, 2017) and the conceptualisation of museum as dream spaces (Kavanagh, 2000) and as repositories of pedagogies of feeling (Witcomb, 2015b) and affects (Waterton and Dittmer, 2014) are particularly relevant in the discussion that follows: the making of Beamish Museum is a making and remaking of the regional imagination.

The dynamic agential nature of memory, with its compelling role in heritage co-production, is interlinked with wider materiality – past and present, material and immaterial (Domanska, 2006). Collective memory is variously tethered to "places, ruins, landscapes, monuments and urban architecture, which – as they are overlain with symbolic associations to past events – play an important role in helping to preserve group memory" (Heller, 2001: 103). We could as easily substitute the term 'heritage' for 'memory' in this sentence, to much the same effect. Herein lies the promise of collaborative heritage practices and museum outreach projects; they allow communities to fill the gaps where memory fails through imaginative, inclusive and open-ended strategies to bring the past to life (Kavanagh, 2000: 4). Communities like Kibblesworth and Ryhope resist being reduced to the stereotype of a monolithic housing type or old-school picture-house. The communities resist this flattening by storytelling – by reaching out with their memories of everyday life that did not begin and end at the mine shaft. The two villages perceive themselves not as victims of deindustrialisation and poverty, but as dynamic social groups who, while striving for a new post-industrial sense of community, have dispersed but are still very much *in place*.

In this sense, social history museums such as Beamish Museum attempt to foreground displays based on visitor experience and articulated in ways that attend to local cultural practices. In the case of Beamish, the current focus is on the imaginaries of the 1950s in the northeast of England – above and beyond the legacy of the coal mines. The 1950s project attempts to bring together the various tetherings and imaginaries that make up sense of place across time nurtured and negotiated by local communities. At Beamish, the exploration and research towards the 1950s town have constituted a journey of discovery where doing together and sharing remembrance can constitute a "shared feeling voyage" (Wetherell, 2012: 77).

Memory inhabits the entanglements of thing-person-place at any given locale, and materiality can lend an added layer of understanding, a depth of feeling to the learning that is a community and a depth to the storytelling neighbourhood of the villages. Materiality can represent a tacit knowledge of the social (Cruikshank, 2006: 9). At the core of my collaboration with Beamish Museum was a desire to build on oral history and memory work to include the materialities, textures and imaginaries of the local 1950s. The value of such a collaboration, recalled in the following, is not so much in the ethics of co-production and co-authorship, but rather in the potential of similar methodologies to open up social histories and memories from below in the public sphere of the museum. To highlight memories means to re-evaluate the meanings that communities ascribe to things and places. It means understanding how people used to live and how communities exist now caught in the ever-present entanglement of past, present and future identities.

In its material and immaterial traces, community memory is tethered to place and is all-pervasive. A learning together of what makes the past meaningful can give direction to a community who may find itself at a loss on how to fashion itself. The multivocal process of heritage value co-production, the learning by doing and by being together, may be a strategy for future shaping (see Dedrick, 2018 for a creative mix method participatory technique: photovoice). Scholars from different disciplines – among others, Alessandro Portelli (1997, 2003), Francesca Cappelletto (2005) and Beverley Butler (2009) – have argued the case for remembering as oral history or archival practice as a dynamic act of assertion of one's position in history, within or outside of ingroups, and the value of such traces as tools for social justice. This ethos is what has guided Beamish Museum in its continuing engagement with working-class memories and stories from below.

But I have to ask: through our conversations, questioning and doing together, have we, the professionals, enabled the co-production of heritage understandings at the local level? This is the place to offer some reflection on our practice. Starting from the premise that the journey of memory and place are fundamentally social, embodied and interactional, this chapter reflects on the benefits of a methodology that intersects engagement with communities, the imagination, mapping experiments and collaboration. I start the shared feeling voyage with a survey of its building blocks: stories. Stories are the foundation of the shared feeling voyage that Geraldine Straker and I compare later in this chapter. The project started with the collection of stories from the community at the museum and out of the museum, in the 'real world'. But at this stage Geraldine Straker and I recall our respective encounters with the memory/heritage mapping methodology; we reflect on the effects and impressions this process has had on our professional and personal sphere as well. It is an ongoing conversation: what follows is the result of many conversations, text messages and email communications but also our reciprocal take on a joint experience. The reflexive interlude section is a dialogue between and among professionals who come to the community mapping method from different backgrounds and with often different expectations and requirements. This exercise attempts to reflect on some more practical but also a more hands-on, unmediated emotional aspect of doing 'memory' fieldwork. I will get the ball rolling by telling the story of how the process began.

Sarah's experience

In October 2015, when I initially approached Beamish Museum about working with them on some 1950s memory map for Remaking Beamish, I had no idea about what things, places and communities I would be coming into contact with. I had just started a post at Durham University and wanted to get my bearings in the local area by shaping the project I was working on to my confirmed strength – the making and enabling of community-led memory mapping. I knew Beamish was planning a huge project named Remaking Beamish, which would involve building an extensive 1950s town. I was excited but also worried. Mainly, I had no clue as to whether the museum would be interested in doing this kind of thing, and I did not know if the communities in question would be up for a mapping of their local sites. I met up with Geraldine Straker and Lisa Peacock of the Engagement and Participation team at Beamish and proposed the mapping strategy to them. I pitched it as a way to contextualise the villages and places where the reconstructed or rebuilt buildings originated from in order to provide background for visitors. In addition to this educational aim, and perhaps even more crucially, the conception of the memory map was to ensure preservation and transmission of some of the place attachments and emotional living memories of the residents and patrons of such places, once they were out of their context and in the new 1950s town.

In order to illustrate the practicalities and possible outcomes of the mapping exercises, I produced three sample community-led maps I had already completed, and both Geraldine and Lisa were very pleased and excited about the potential outcomes for the museum exhibits in the upcoming 1950s town. At our initial meeting, I was able to view plans and sketches for the forthcoming Remaking Beamish project – at the time the scheme was in its development phase and had secured stage one support from the Heritage Lottery Fund.

This kind of memory map-making was as new to me, in a sense, as it was to Lisa and Geraldine. Previously, I had worked towards heritage site–style maps that encompassed larger areas, such as historic landscapes and whole neighbourhoods in towns. This new way of thinking and representing the memories of buildings and structures with a surrounding affective economy of homes, shops and landmarks was new to me. The idea of working with communities to pin down and visualise where people lived, where they went to school, where they had a drink, where they spent their leisure time, was a fascinating prospect.

From the start, I had some concerns of an ethical and positionality-related nature, however. As an academic, how I position myself in relation to my research and my co-researchers is of vital importance. Previously, I had worked with communities of which, in a way, I was a part: I had traced the histories and memories of a town, a region I knew or had lived in. In other cases, I had worked with and within communities of memory of an 'ideological' and emotional connotation: Resistance fighters and their descendants, which whom my family could identify. I had some stakes in the research, and I knew I could make my voice heard as the ultimate editor and producer of the memory maps in question.

On the other hand, in the northeast, an area I barely knew at the time, I knew few people – and mostly in the academic context of Durham University. In the greater County Durham, I could not claim membership or even an understanding of the local culture, of the histories and experiences of former coal mining communities. I felt like an outsider imposing a methodology on local residents. The idea of being perceived as 'yet another person from the university' troubled me because the idea behind these maps is to encourage spontaneous participation by all and to work together to visualise or represent memories of a past that is still acutely felt and relevant to the present day. In order to reach this spontaneous engagement and openness, there is need for a rapport to be established.

But I was lucky. Thanks to the friendly and helpful mediation and availability of both Geraldine and Lisa, I was able to integrate seamlessly with two wonderful communities of residents, one in Kibblesworth and one in Ryhope. My encounter with the people who lived in, or had worked at either location, was amiable, good-humoured and often downright uplifting. The local co-researchers welcomed my involvement in the oral history collection for the 1950s project and were fundamental to the shaping of the idea. The community members' and Geraldine's insights and often inspired ideas about the maps re-energised a methodology I had perhaps started to take for granted. Geraldine in particular injected so much enthusiasm into the project that I started to look at this map-making process with fresh eyes. She circulated and brought the map into community meetings even when I was unable to attend, effectively adopting the 'visualising' experience as an engagement method – a professional satisfaction which always occurs whenever a museum or community member 'takes to' the method and implements it.

The exercise of sticking sticky notes to a large paper version of the map has meant that community members engage with the memory mapping process quite informally and in a fluid, spontaneous manner – sticky notes can be easily replaced, moved, taken off altogether or stacked up over a particular spot. Both

Geraldine and I were, and still are, aware of the limitation of the paper medium in terms of interactivity, as only a number of people access and interact with any given map at a time, but this is a weakness we can work with and learn from during any future digitisation phase of the Ryhope maps. After talking with Geraldine, the idea of making an accretional deep map public, enabling people who lived or have a connection with Ryhope and the Grand Electric to make their mark on it, has rapidly become the solution to the 'segmented' nature of paper-based map interactions we had in Ryhope or at the museum.

The idea of working with communities on their memories, making their maps, is exhilarating. It means we can co-produce knowledge in terms that make the community feel vindicated, present and in charge of their vision of the past, present and future. These visualisations map futures, expectations and hopes, not just things that are gone or over. In working with groups at Ryhope, for example, the emphasis was on the pride for work well done by the former cinema projectionists to show the wider society that the northeast was not just about the pits. In Kibblesworth, people felt privileged to be housed in a dwelling with inside toilets. These details, albeit not explicitly detailed in history books, matter. That is why I was so excited to be working with staff from a museum of the People – not just of the Past.

I believe the main way in which this joint methodology has benefited Beamish Museum, the community and myself as researcher has been the recognition of a making-by-learning. A learning about each other's expectations, memories and understandings of place. In the end, the early drafts of the maps of Ryhope and Kibblesworth are just that – early drafts, early steps in an ongoing conversation which, I hope, will prove a pleasant and stimulating experience for the communities involved. Over to Geraldine now.

Geraldine's experience

When Sarah first approached the museum, I wasn't quite sure what to expect. The first meeting was really interesting, though. The deep mapping described by Sarah seemed a natural thing for the museum to be involved with and to try as part of the community and development work we were undertaking as part of Remaking Beamish. The use of maps with groups is not unfamiliar to the museum, particularly for example working with younger people to explore how their community used to look and to identify the changes and developments that have taken place across time to the present day. However, this work tended to be more about exploring change than creating maps.

Geography is important in terms of the museum's work. Community participation is often based around geographical communities: the locations of historical events, areas connected to particular objects the museum has or the places where buildings have been (or will be) moved 'brick by brick' to the museum or where buildings are to be replicated from.

Deep mapping as described by Sarah seemed to provide another way we could connect with particular communities that would result in a visual representation of memories and connections to place. Rather than just recording memories, it would provide a visual representation of them and show just how they were connected to place.

There was instantly a lot of scope and potential seen in this way of working that would enable us to work with and involve people in the project, to listen to and learn from memories and gain a better understanding of the places where Beamish intends to move buildings to or replicate buildings from.

The ability to work within 'living memory' and to create a time period at the museum that is familiar to people is an incredibly powerful and important thing. Through previous projects we are aware of the connection and pride that can be developed through the museum exploring local history with people, connecting generations and then celebrating aspects of that history both in the community and at the museum.

As a starting point in the development phase of Remaking Beamish, deep mapping provided an opportunity to work with and start to build relationships in communities connected to the plans of the museum. The maps as ongoing documents would enable people to input their memories and reflect on or comment on other people's memories. When on display maps provide information to other people who may not be familiar with an area. As a museum, we look at everyday life in the northeast across different time periods. The maps would essentially be a map of everyday life experiences in a particular area, where people have lived, worked or gone to school. The mapping we have done with groups has been an interesting and engaging task. People have eagerly shared and taken part in sessions, wanting to share their memories. Essentially, we have done it in three ways:

1   At a group event where individuals came to look at the map and talk to Sarah or a staff member and talk about and point out places and memories, largely on an independent basis
2   As part of a group session where other activities have taken place and where people have come up in small groups to look at and add their own story or memories onto the map
3   Not necessarily with the map in the forefront of our minds, but through working with individuals who offered stories, such as oral histories or other face-to-face meetings, we found we could potentially add these stories to the maps at a later stage

Mainly the sessions we did were with a medium-sized group who came up in threes or fours to look at and add their own stories or other people's stories to the map. You could see a great deal of interaction between people with memories from a similar time – people working together to remember names and share stories and experiences. Occasionally this can lead to discussions over memories, particularly over where things were, and a continuous refining of thoughts. It was good to keep involved in the discussions to hear memories but also to help provide some balance and ensure everyone was equally able to input. We did find that often people were more interested in vocalising their memories, not wanting to personally write them down – but happy for the facilitator to do so. The conversations often continued after leaving the map, and sometimes people wanted to come up again.

Practically only a limited number of people can access and work around the map at any one time. You could have individual maps, but this would lose an important social and sharing aspect. The first way we did the map also lacks an interactive element. The map itself can also quite quickly become full of information; the use of sticky notes or labels numbered to correspond with numbers on the map helped. Keeping track of these can also be a challenge.

From the point of view of our current project, Remaking Beamish, we were particularly interested in 1950s memories to understand and build up a picture of a particular community's memories and experiences during that decade. As with all our work, though, we're mindful that people don't necessarily remember things by decade, and there is value having that depth of understanding about an area across different decades. Sometimes it was helpful to understand when the memory stems from – particularly if lots of people are inputting information, as of course places change over time.

The mapping process has certainly been interesting from a learning point of view of understanding the place we are working in. It has also provided an interesting way to engage people with the project. We very much consider the map a working document that we hope to add to through the project. As well as this being a group activity, we have found that working with individuals has also provided content for the maps. Through oral histories with some people we have heard wonderful stories that can also be added to the map. What is interesting here is that because the primary activity has not been the map as a group exercise, we have heard memories that focus on experiences that can then be mapped.

Currently we have only worked with the map with people with older memories of the location as well. I think it would be really interesting to explore the map with younger people to see their responses to the comments already added – but to add their own thoughts and places that are important to them. Are the social physical places people go to the same across the decades? Are younger people aware of the physical changes that have taken place in their community, particularly thinking about substantial changes that have occurred since the closure of the local pits (Fieldhouse and Hollywood, 1999)?

Around the same time that we started working with Sarah on the maps, Beamish was fortunate to have a creative writer in residence working at the museum. This writer was also working in some of the same places as Sarah. This intersection has brought some interesting possibilities to light. Not only could people add 'facts' to the map but through the writer's work and workshop ideas sense could be evoked through simple writing exercises. These memories, thinking about sights, smells and sounds of a particular place (in this instance the cinema), could also potentially be mapped.

The writer's other work during the residency included a writing day at the museum, and this led to the idea that you could not only create a deep map of memories and experiences but if you wanted to you could also create a map of creative responses to places as well. Through Becci's work (the writer in residence) we received some creative writings about places in the northeast that could also be mapped. Both elements have currently been paper-based exercises with people. While we are aware of open-source mapping programmes, this is not something we have taken forwards yet. I think with some groups the nature of being able to sit down with a cup of tea and share, listen and talk to each other and the facilitator about their memories is so important. It is relaxed and informal and easily accessible.

As discussed prior, people were often more at ease with vocalising their stories rather than wanting to write them down as well. Inputting individual or collective stories and the augmented, scribbled-on maps in a digital format would be a step further away and would require extra resources. Sarah and I were aware of this potential development during our meetings and agreed it would be the next logical step. At present this activity is not precisely planned for, but this is not to say someone could not later input information into a computer – as Sarah has done for early drafts of the Ryhope and Kibblesworth maps.

Remaking Beamish is now in its delivery phase, and we hope to continue the work started with Sarah. As we deconstruct the cinema in Ryhope and rebuild it at Beamish, we intend to work with a number of different groups and individuals in a variety of ways. I would hope the Ryhope memory map is something we can continue to work on as the project develops. The mapping process may also be something we can look to do in other areas the museum will be working in.

A shared archaeology of the dream space

Both Geraldine and I have reflected on the practice of community memory mapping, its limitations and its potential for social learning and peer exchange. Rather than memory work, Geraldine and I may possibly have operated on a level that identifies and stimulates the foregrounding of emplaced postmemories, bringing to the fore the living and loving connections between generations of residents and their families (see also Field, 2014). We may have operated within a local and regional economy of caring (Ahmed, 2004), a framework for the exchange and circulation of cultural and affectual meanings and intentions (McCormack, 2003; Wetherell, 2012) destined to animate the new museum display of the 1950s town at Beamish.

In asking what benefits such a project has to offer to researcher and museum staff, we may reflect on the nature of the storytelling enacted through this methodology. In its focus on stories alongside material culture items, was this experiment a form of visual oral history or an archaeology of the senses? Does the attention to the material and affectual bring this research into the realm of the antiquarian's imaginary of a past grounded in emotional understandings? Where does the community input end and our (expert, trained, professional) reshaping begin? At any rate, the affectual and the scientific aspects of the Remaking Beamish project need not contradict one another. The intersection of community values and professional engagement is worthy of dedicated and proactive agendas (Matthews and McDavid, 2012). Community-led and community-shaped archaeological and heritage practices are as unpredictable as they are powerful – this synergy between locals, activist groups and the 'past specialists' shapes and gives meaning to events, places and things, big and small (McDavid, 2010; McAtackney, 2015). I would argue that, in the context of the logics of heritage co-production, memory-making from below can be achieved through a simple, inclusive methodology that listens to what people have to say (McDavid, 2002; Kiddey, 2017). I propose that the joint mapping method be tried whenever there is a wish to gather together multiple perspectives on place and identity or in contexts affected by competing materialities and narratives (Sobers, 2017).

Similar methods may be a starting point to articulate some of the many ways to work with communities on their remembered, imagined or buried pasts. Geraldine's reflexive input has highlighted the benefits and potential ways of improving the interaction and engagement of the community map-makers. In the context of a social history museum such as Beamish, the potential educational and empowering value and place of such maps is becoming increasingly clear, as is the museum's role in creating shared visualisations of a past that is very much present in the communities today (Annis, 1986). A web larger than the Grand Electric itself, the memory mapping exercise has brought together generations of residents, professionals and friends of and in Ryhope in an endeavour to do justice to the village and its vibrant sense of place. As Geraldine notes, the Ryhope memory map is "something we can continue to work on as the project develops. The mapping process may also be something we can look to do in other areas the museum will be working in". This makes me happy, the thought that we did something which may be meaningful to communities as they visit the museum. The visualisations which the communities in the northeast of England are creating live in a performative space in between private memory and public consumption of the past (Crouch, 2010): the main thing is, this process needs to happen collaboratively, without the exclusion or elision of local views (see also Dedrick, 2018).

Whose dream space?

A focus on the local, the familiar and the vernacular is at the core of heritage collaborations. Aside from powerful exceptions in studies of 'other' heritage-making or meanings (Hall, 2005; Meskell, 2003; Field, 2008; Koskinen-Koivisto, 2016; Kiddey, 2014; Kiddey and Schofield, 2011), heritage is still usually framed within the ontology of the known, the within reach, the approachable, the ownable. Even the liminal is a pedagogical space we can make sense of: writing on the liminal space of the museum, liminality and learning, Puar frames these as "events, actions and encounters between bodies, rather than simply entities and attributes of subjects" (Puar, 2012: 58). This liminal but grounded mode of encounter affected most of the case studies I have articulated here, aside from the sinkhole in Chapter 5, which, while liminal, is neither ownable nor easily learnable.

In comparing my experience with Geraldine's, I am reminded of Stewart's idea of ordinary affect as a comfortable place, a well-known and familiar thing (2007).

The ordinary can channel such powerful associations and stories, leading to the production of pervasive and influential events and strategies for existing – for *being*. Esther's yet-to-become reconstructed home, at the time of writing, already resonated with local affectual connections and sense of place; it is an ordinary house: a typical semi-detached dwelling evoking all kinds of associations and homeliness in those who will visit and imagine Esther there, or their grandmother or both. The affectual tethering will likely always be there, remembered, conjured up, stimulated by being emplaced together in a homely space, if only for a few minutes. Despite the liminal nature of the whole 1950s village and the potential uncanniness of this particular dwelling, visitors will recount their experience framing it in the familiar: "it was just like being in granny's house", some will say. And yet Esther's Sunderland semi-detached dwelling will always remain a liminal place: no longer there, not quite 'here'.

Affects are not just the ephemeral ingredients of stories and experiences: they are a sensory and more-than-sensory glue, a paradigm for a community as a way of learning. They are shared things that are felt, and possibly learned; shared, and let go of. This process of sharing goes beyond the evocation of memories but may well lead to a concurrent construction of contemporary and future imaginings and understanding of the social (Navaro-Yashin, 2012). I am also mindful of Kavanagh's admonition about restricting the museum experience as a dream space to memory: "it is inappropriate to confine ourselves solely to this if we wish to understand dream spaces. Memory does not exist in an emotional vacuum nor in a space devoid of the irrational" (2000: 9). The implication for our collaboration here is that the reticulation of the dreamlike, the imagined, the remembered and the idealised interpretation of the local 1950s has emerged, and is still constantly emerging, through affectual manifestations of heritage; through participation, through a doing together; through imagining together and drawing maps that pin down stories and memories.

Archival commitment

The museum's "archival commitment" (Kavanagh, 2000: 3) is such that the narrative it fashions about the 1950s has to be born of local stories, collected with communities. The museum itself becomes a storyteller among storytellers just as I am a map-maker among many map-makers. We share a story: the academic, the local communities in the northeast and Beamish Museum: we share the same "dream space". A shared story offers a foundation for a shared identity, a moving forward with others. Whenever the past is constructed as a *story*, memories are "prepared, planned and rehearsed socially and individually" (Schudson, 1995: 359). Schudson goes further: he proposes that collective remembering entails some extent of instrumentalization geared towards a current strategic end. In the case of the community of Ryhope, the current end seemed to coincide with a desire to capture and communicate the lively memory of a *village*, in contrast to the extrapolation of one *building* to Beamish. This foregrounds a translated dream space, a journey of affects from place to place.

The insights and experience of Geraldine Straker in this chapter lend a more nuanced insight into the other end of the encounter with the mapping methodology, seen in this case through the lens of the museum professional. Geraldine's invaluable input and support in facilitating my introduction to our northeast co-researchers in Ryhope and Kibblesworth have been crucial in my research. She, alongside Helen Barker and Lisa Peacock, was encouraging from the very start, proposing that we include the 'memory maps' in the Beamish Museum's successful Heritage Lottery Fund bid for the 1950s extension. I will be forever grateful for the team's enthusiasm about trying out this fun, if slightly unconventional, way of opening up oral histories and storytelling. I admire Geraldine's immediate confidence and creative insight in initiating mapping exercises, eliciting information from co-researchers through maps and making the process enjoyable for participants. I can only commend the effortless way she has been working the insightful 'everyday' stories she collects into dynamic and brilliant spatial and pictorial stories. I could not have hoped for a better partner at Beamish Museum.

In actual terms, the 1950s town at Beamish Museum will become an emplaced postmemory in its imaginative inception; postmemory's connection to the past is "not actually mediated by recall but by imaginative investment, projection and creation" (Hirsch, 2008: 107). I wonder to what extent our fieldwork has allowed or enabled elements linked to imaginative investments in Ryhope to emerge through storytelling and map-making. I wonder if we succeeded in bringing to the fore the 'little' memories that matter. Coming face to face with feelings of loss, or the acknowledgement of a longing for a past now gone, in its many inceptions and manifestations, is something we could and should embrace and work with. We should not try to dismiss nostalgia or try to expunge it from our outputs as uninteresting or irrelevant. Any exercise in participatory research and memory-making does not operate in isolation from the current socio-economic expectations and circumstances of communities that have been deindustrialised (Walkerdine, 2009). These concerns grow in tandem with current political discourses and wider identity politics.

Have the visualisations served as a creative projection, a linkage between past and present (and leading to an exhibit/heritage afterlife?) If some projection was among the current aims of the residents of Ryhope, we embraced it wholeheartedly. We took projection and transfer as starting points, even: that is, at least, what I think we did. We foregrounded the memories and imaginaries the communities chose to share and construct: we left it up to them. While we know that memory is always memory of an intersubjective past lived in relation to other people (Misztal, 2003: 6), we had to figure out ways to pin down these memories to render them visible, vibrant, intuitive and easily communicable and transmissible to others.

Heritage values visualisation, such as the 1950s project at the Beamish Museum, which has affected us as the practitioners in similar ways as it has touched the lives of our local co-researchers. The project has made us question our place in the research and indeed in the experience of these cultural worlds (Classen and Howes, 2006). The space of the social history museum is an affect-rich domain where visitors "discover unsuspected links to each other and are made aware of their own narratives about themselves and those around them" (Witcomb, 2015a: 164). This process also involves the enablers, collectors and sharers of stories and memories. After a series of engagements and conversations, we have produced a pool of knowledge about what the Grand Electric was, is or will be. Our participants have some of the information, and the museum holds some more. The whole assemblage of knowledges in which the maps are plugged tells the stories of one (or more?) place through multiple perspectives. All the data and information any of us, participants, residents, projectionists, curators, hold or possess is partial, incomplete; as with all qualitative research subjects, the building's story, its materiality, is always in a process of retelling and re-remembering.

Expanding on these ideas, I argue that our joint engagement with communities of memory serves to situate heritage co-production (together with residents and heritage publics) as affectual moment of encounter, a process embodied in the practitioners' own experience and relational practices (Mulhearn, 2008; Dedrick, 2018). That is, our relative success or failure in engaging the community of Ryhope was dependent on the reactions, interactions and responses of others. Tacit knowledge is "embodied in life experiences" (Cruikshank, 2006: 9). A different kind of knowledge to that usually housed in museums, but perhaps more pervasive, rooted in affect (Sather-Wagstaff, 2011). Community may be working as a way of learning from within, reaching out: affectual and emotional experiences in the heritage process thus erupt from the biographical, historical and social time we are woven into (Macdonald, 2013; Waterton and Watson, 2015b). With this in mind, the following chapter attempts to offer not a conclusion, but rather a summative reflection on the ideas and practices explored in this volume.

8 Moving forward
Not a conclusion chapter

The afterlives of stories

The mapping experiments varied across the case studies and cross through time and place. So, for instance, avocational archaeologists and historians lovingly navigated a hillside in northern Italy wishing to get closer to the secrets of their ancestors; this community's heritage practices are bound to the land in a respect-ful yet informal bond of stewardship. Former mining communities in the north-east of England anchored memory to often run-down former pit villages, wishing time to slow down or to accelerate into a fairer future but also impatient to see their social values represented (see also Orange, 2015). The consciences of Italian Fascists and anti-Fascists both wanted time to rewind in order to be able to fore-stall the perpetration of violence, wishing for the fabric of their town to become whole again. What afterlives do such stories possess? What traces do the stories leave behind, and to whom are these traces visible? In each case, the ensuing map-ping fieldwork at least attempted to listen to and respect the community's wishes; the maps have tried, if anything, to visualise local visions of place as best as can be hoped: with openness and collaborative spirit.

The focus of this book, and of the strategies therein, is on the local processes through which people deploy affective, active, imaginative resources to connect up with the past – not as history, which is remote or belongs to someone else, but as directly experienced and remembered. A focus on the experiential breaks down dualisms of culture/nature, inside/outside through the holistic perception of multiple timescales and spatial dimensions contemporaneously. For Ricoeur, the past is all around us: a thing of the present. "As a physical entity the trace is something of the present. Traces of the past exist now: they are remnants to the extent that they are *still* there, while the past context of the trace – people, institu-tions, actions, passions – no longer exists" (1991: 345). This idea is fundamental to the memory-map projects adopted in this volume where I have shown various potential applications of a simple, easy-to-do and inclusive methodology to gather stories in a colourful setting. Therefore, the 'memory maps', or community maps or deep maps – does the name matter? – may constitute a visual and interactive way to conjure up pervasive traces of the past.

Overall, the maps and visualisations we encountered may attract, foster or stimulate communities' interest or curiosity, with a collaborative mapping medium spearheading participation in heritage and history-making. The visualisation or mapping process serves to channel, or 'bring to life', stories which sometimes risk remaining untold due to their ephemeral, controversial or simply 'mundane' and homely nature. Calvino's protestation that we can have too many stories resonates to those of us overwhelmed by a multitude of voices telling stories of the past and present – and when we instead perhaps hope for specific outcomes. We are overwhelmed when we do not know how to best harness stories: spatial stories, object stories and people's stories intermingle with inventions, imaginings, myths, even gossip and rumour in the thick, multivocal experience of everyday life and everyday pasts (Jarrett, 2013). This chapter does not presume to provide closure or a conclusion to the processes set in motion by people coming together and 'doing together' (Keller, 2003). Through gathering sources and thinking about place, communities and individuals may materialise memory and sense of place through collage-like visualisations, or they might choose to work differently, through different channels and engagements.

Open-endedness is part of the process, and subsequently this part of the volume does not attempt to bring these explorations and reflections to a tidy end: we might instead reflect on some key aspects of what we encountered thus far and make sense of some of the advantages and potential pitfalls inbuilt in one methodology among many more possible strategies. As versatile and eclectic as the varying dimensions of memory and place attachment of various communities, the mapping methodology has served as a connecting tissue gathering together strands of perceptions, projects and pride of place. It may have had different outcomes in different communities, but we are not yet to know unless (until?) we propose it to other communities. At the time of writing I am initiating a similar story-map project with two communities of asylum seekers in Italy, for whom place-making has a very different meaning. We shall see how the experiment works with these displaced co-researchers. In the meantime, what more can be said about this method? There are variations on a common theme, which is a process where a researcher or practitioner moves around with a group of people, then asks them to get together at a later stage to 'percolate' their feelings and emotions in the real world onto paper or a computer screen. It may now be helpful to briefly revisit what the mapping and visualisation process entailed in each of the case studies described in the volume – by way of summing up results, so to speak.

I have chosen the five case studies – Monte Altare, Vittorio Veneto, Bus de la Lum, Ryhope and Kibblesworth – as they each engage with the notions of place, memory and the imagined in different ways, but in ways that are connected up through a complex set of behaviours and stances. If I were to be pedantic, I might suggest that some among the case studies confront the three different 'ideas' encapsulated in the book's title more poignantly than the others. So, for instance, the Monte Altare and Bus de la Lum projects have more to do with the imagination of place, of deep time and of current perceptions of heritage objects as imaginative materialities (Monte Altare) or in relation to ongoing mythologies of wartime massacres (Bus de la Lum). We might say that the Vittorio Veneto experiments and the Ryhope and Kibblesworth projects are more explicitly about memory – melancholic and empowering in both cases. But things do not work that way: in any case, all examples express the ways in which communities or individuals construct place and make sense of the past in the present through memory, place and the imagination.

The projects I introduced have addressed a deep-time memory experience of a longue durée landscape (Chapter 3), a town's experience of absent-present traces of a violent past during the Second World War (Chapter 4), a non-place and its imaginings (Chapter 5) and the pride of place of two communities in the 'deindustrialised' northeast of England (Chapter 6). To an extent, all place-specific mapping experiments represent ruptures from traditional practice in historiographical and archaeological ontologies and frames of reference. They do so in vastly different ways. In Italy, communities contest and defy the predominant scholarly and mnemonic tenets, and their challenge disrupts mainstream heritage representation. Meanwhile in Britain, two villages in the former mining regions of northeast England have crafted a mnemonic reality that looks past often 'embellished' nostalgia practices and ultimately harbours a deep cross-generational sense of community-centred optimism and solidarity.

The Monte Altare

Mapping heritage experience through the senses explores people's engagement with heritage sites and opens up places as sites for discussion, rather than occluding them as 'off limits' and restricted to elitist datasets of artefact typologies and the like. The simple, collaborative mapping technique can bring to the surface the various spatio-temporal layers of places and things with historical depth: heritage as experienced by stakeholders, then, becomes part of an ongoing negotiation. This was particularly useful in the case of the Monte Altare: the fieldwork map in Chapter 3 reminds us that community fieldwork can be fun as well as a challenging and stimulating way of learning about the past together with other like-minded non-professional actors and scholars. The main challenge in this case was to produce maps that put community insights at the centre of the process while also engaging with archaeological materials and historical data in a way that would satisfy the requirements of a doctoral thesis (mine!). The main drive for the community was the establishment of their unique sense of place as the dominant narrative in the face of a neglected publication record by institutions in the region, which despoiled the archaeological site of Monte Altare of its rightful community and 'sense of place'. To people living near a multi-period heritage site like the Monte Altare, for instance, there can be no pre-emptive division of time periods or separation of nature from culture, as the landscape and site open up as a whole in experience. The mapping exercise started in the field thanks to the guidance and insights of my non-professional colleagues and interested residents. Their sense of place was paramount to my interpretation of the archaeology: I would have felt

lost without the knowledges and celebrations of the (deep and recent) past performed in the field by citizen experts like Giorgio Arnosti, Carlo Forin and other
Gruppo Archeologico del Cenedese (GAC) members and affiliates.

Collaborative maps such as this also position heritage understandings as more-
than-temporal: they become situated and embodied and enable self-projection on
the part of communities – all of which are central to heritage co-production. More
than metadata in the margins of a document, fieldwork stories tell us about finding
and excavating objects and trace the changing lives of things and places through
time; the heritage items, understood within the stories, shaped understandings of
this site. Stories make a site unique, in the same way that sites are assemblages
of the stories told about them. The local citizen experts' insider perspectives did
not just influence my analysis in the thesis – their stories gave life to the thesis, or
better, they brought the thesis to life.

The Vittorio Veneto civil war maps

In line with recent thinking on grassroots heritage (Kiddey, 2014, 2017) and collaborative mapping, heritage professionals and communities are working together
to map out and express what heritage 'feels like' in the everyday (see Onciul et al.,
2017). This participation drive can be extended to dark periods of social history
such as the Italian civil war of 1943–1945, when Italian Fascists fought, injured
and killed Italian anti-Fascists and vice versa. Qualitative 'deep' mapping can
effectively contextualise the violence and hardship of the Italian civil war in order
to untangle the elusive mechanisms of othering, exclusion, alienation and violence that animated those times. Thus, the somewhat sombre and decidedly divisive memory maps of the civil war in Vittorio Veneto trace ghostly geographies of
danger and violence. Doing research here, with this community, was never going
to be pleasant: the most obvious issue at play in that context was the sensitive and
harrowing nature of the research itself. I was in the midst of a community aware
of a storyline of unpunished (unpunishable?) crimes that still cast shadows onto
the present-day affectual geography and yet, at the same time, a geography which
was muted and invisible to most. Collectively, we feel that unearthing memories
on the war-scape maps helped bring to the fore, or resurface, marginalised stories
of pain, betrayal and violence that had mostly seeped through the cracks of pave-
ments, lurked in street corners and still throbbed and ached in veterans' bodies
(cf. Macdonald, 1997). The imaginaries of what happened, what should have happened and what should not have happened haunted fieldwork from preparation to
map-making process.

We looked for a way to make room for the pain and social division in our visualisations of the past, which also encompass present understandings and a lesson
for a fairer future. And yet my colleagues and I made the logistical decisions to
maintain a degree of separation between those supporting the partisan cause and
those who aligned with the right-wing faction in the town. For obvious reasons,
we did not want to create unnecessary discomfort or stoke any further disagreement and enmity in the community. The desire to keep the two sets of 'memories'

separate during our walks was, as can be imagined, a barrier to a truly holistic experience of place. That said, the younger participants were the most vocal advocates of a 'full' encounter, and their intervention may yet prove beneficial in the years to come. At the time, however, people walked in groups that tended to agree on 'what happened' and who the victims were. While a deeper confrontation lacked, we also avoided overt unpleasantness. With all groups, we made the decision to stage all phases of the design and display of the maps after fieldwork as an open participatory process. This was a step towards bipartisan inclusivity: an attempt not sufficiently developed to 'revolutionise' the memories of the war, but a step forward nonetheless. At the time of writing, this public engagement with our walking memory maps has not yet taken place. When this happens, however, the entire community will be able to drop in at the local library and have their say, irrespective of whether they actively participated in fieldwork or not.

The Bus de la Lum

In my work on the postmemory of the sinkhole in the Cansiglio forest, I encountered traces of a past that are both haunting and unmanageable, there and not-there. The spectral embodied *and* body-less geographies traced by the bones of the victims, and the cross unofficially erected on the site of Bus de la Lum to mark a 'massacre', taunt our conscience and test our human compassion. If, as has been suggested (Stewart, 1997: 125; Hillman and Mazzio, 1997), the human body is a way of relating to and perceiving the world in its real and imagined dimensions – then the fragmentation of corpses, artefacts and landscape at Bus de la Lum joined together to fragment reality and make this place something else, at the edge of memory.

Visualising this non-place has not been easy. Shifting and ambiguous shades of collaboration, collusion, culpability and violence in the Cansiglio area had been left largely unstirred, living at the edge of speech, confined to the intimate realm of the family lore and embedded in the somatised memories of violence and of the victims. This part of the project was, overall, the most distressing participatory experience of my career; in all honesty, my encounter with the communities of Cansiglio and the veterans left me emotionally drained. Much unpleasantness emerged during our talks and hovered in the conference room of the hotel where we initially met, and this experience affected many of us, our families and our allies. The most critical challenge in this exercise was the need to be respectful to all views and emotions being expressed without preconceptions. I have argued elsewhere (De Nardi, 2015, 2016) that such an impartial perspective is sometimes hard to achieve, especially when a potent autobiographical element becomes part of the story/stories being crafted and lived through. While about a third of co-researchers were happy to be named and consented to the publication of their names alongside their quotes on the maps and in related literature, most asked to remain anonymous. The general feeling was that these participants would make their mark on Bus de La Lum simply by talking about it – which had not been officially attempted before. It seems as if the wish to bring some degree of agreement and coherence to perceptions of that chaotic site was sufficient to make the whole exercise worthwhile and to appease participants' uneasy awareness of the place. On reflection I am convinced that it was better to attempt to visualise Bus de La Lum than to leave it alone; easier to face its demons than to leave its occluded memory languishing at the edges of discourse. If anything, openly discussing what the place means to communities living in the immediate proximity of the site has served to lay some of its unsavoury ghosts to rest.

Beamish Museum and the English northeast

The northeast maps may be seen to serve two purposes: the first, endorsed by the museum in the first place, was to generate excitement about the 1950s extension at Beamish. The second was the drive by the communities where some of the buildings once stood to ensure the relocation of a whole 'community of memory' to the museum, with the aim of maintaining a sense of home and leisure in those communities beyond the extrapolated building themselves. Through memory mapping, both Geraldine Straker and I initiated a dialogic reflection on the practice of community memory mapping, with its limitations and its potential for social learning, collective togetherness and peer exchange. While any one draft of the maps is finite yet transient, in that more and richer drafts will follow, the subject matter with which we work is also ever shifting and future-leaning. We attempted to 'capture' present and absent traces of memories and identities in the villages, aware of these buildings' transient nature in the landscape. The affectual heritage tethering the cinema and the prefabricated dwellings to the mnemonic landscapes of which they were once part are more powerful than the Beamish relocation and reconstruction rituals: these affects remain part of the local fabric. The verb 'remain', incidentally, reflects a past tense of passivity which does not do justice to the sense of attachment to and ownership of these structures in the collective imaginary.

The peculiarity of the work-in-progress at Beamish and with the communities of northeast England lies in its transitory and changeable nature, apt to be rewritten, rehashed and complicated as more co-researchers grant access to their memories and information. We mapped present, absent and in-between places. In the case of the Aireys, the structures are no longer there in the village of Kibblesworth, but not yet in their allocated space at Beamish Museum. The prefabricated dwellings' slow ruination had become part of the affectual fabric of home at Kibblesworth, a configuration which cannot be evoked again in their new location. In particular, we do not yet know what shape (in a non-literal sense) Esther's home will take. The fact that we have not yet commenced mapping for Esther's home has to do with practical factors and not conceptual decisions on my part. The owner of the house is elderly, and it has thus seemed inappropriate and exploitative to bother her with unnecessary questions while the museum prepares to recreate her home but cannot yet provide the lady with concrete information as to the how and when. As soon as the museum can inform Esther and her family of the practicalities of the re-staging and construction process, we will get involved with mapping her experience of her house's doppelganger going up at Beamish. For now, we can try to imagine what the place is going to look and *feel* like, although we are not sure of either; not knowing is, in a way, part of the learning experience.

In the introduction, I quoted Barbour's observation that "intersubjectivity, countertransference, and the dynamically interacting forces of projective identification, introjection and incorporation [. . .] dissolve the boundaries of self, and are 'other'. They are difficult to teach effectively, and have to be learned experientially" (Barbour, 2016: 96). Learning experientially is at the core of the mapping and visualisation exercises (Cross, 2012): the maps are a learning-by-making. In other words, experiential and *experimental* mapping has a disarming simplicity of its own accord, but at the same time the method entails a potential to achieve multiple goals and purposes. It can be playful or painful, liberating or introspective, and the actors and publics it serves can range from small groups of guarded memory gatekeepers to entire town and regional residential clusters. The potential benefit of the maps and visualisations is in helping us think through memory in a thicker, less isolated manner, as the maps encourage the 'plugging in' of multiple versions of a story. A story may encompass a historical event, a fiction, a private memory, a collective version of events or a chosen official representation, and that story can be a partly or entirely imaginary one.

Space and place are fluid concepts, and each "can refer both to external and internalised conditions" (Lieberman, 2015: 87). Imagination has been a key component or element in the processes and experiments explored in this book. I have made the case for the consideration of the imaginary and the mythological as constituting workable and instructive subject matter in humanities research. The more than real, hyperreal, made up and fantastic can well be, I have argued, a fundamental component of affectual and inclusive understandings of grassroots heritage, cultural or ethnographic fieldwork and so on. The 'real' (whatever that may be) and the invented or imagined all coalesce around places, things, communities. Their *connective tissue*? Stories. Stories have been at the forefront of all experiments and visualisations encountered here, growing out of people's engagement with place and in turn shaping people's interaction with places, things and others (Andrews et al., 2006). We discover stories inhabiting the same affectual spaces as people, time, place. Stories matter to people in the everyday. Real or imagined, stories give meaning to things, to places, to the past and to other stories (Price, 2010). Stories and memories possess a materiality all of their own, which we can engage with and visualise.

Even absent things, even traces of 'gone' things, possess a materiality through storytelling. Thanks to experiments in visualisation and imaginative mapping, communities of memory can trace these elements and make a mark in place and time through pictures and gestures and words of things, people, events and places that are present and absent, visible or erased (De Nardi, 2014a). In some cases, the maps deconstruct why one story is told while another is silent (Alcoff, 1991). The visualisations show the folds within, the memories within the memories and the spaces in between memory and forgetting. And yet the acts of mapping, picturing the world and producing text are, of course, processes belonging to the representational canon (Perkins, 2004; Pinder, 2005). Throughout this book I have argued that experiential memory and heritage mapping visualise and 'thicken' people's sense of the past in ways that are more than representational but not at odds with representation per se – as ways to differently picture history, remembrance, heritage. The maps deconstruct and playfully open up possibilities for shared pasts. I have explored this interaction between 'ways' of sensing the past through five real-world examples that conjure up and blend informal knowledge, ephemeral and intangible heritage and popular history.

Revisiting the maps: a critical reflection

The maps in the present volume do not so much rewrite forgotten or marginalised histories as give space and make room for unofficial and often unwelcome understandings of the past, which are felt, remembered or imagined *in place*. These sometimes playful and sometimes painful visualisations bring 'other' stories to light as vibrant 'objects of concern' in the present – as passionate endeavours of communities who remember and shape their identities through a present-past engagement. Whether this engagement is deliberate or not, whether it feels painful or playful – this methodology has the potential to respectfully and inclusively probe into what it means to 'feel' the past as present in the everyday.

The visualisations made with the communities in this book encompass private and even painful understandings of past events that disclose and re-enact local understandings and versions of events. Some maps hide and recoil from scrutiny, and these are the result of touching private acts of remembrance and catharsis (the maps in Chapters 4 and 5); other visualisations express proud statements of identity and skill (the two maps in Chapter 6). Maps are active agents shaped by people and things while shaping the world in turn. Thus, the maps may be thought as the act of sharing values and messages that matter to communities while also acknowledging things that the community would rather keep silent about. To reimagine a story means to give new energy to the past and to revitalise the present.

Advantages of the mapping experiments

The process of co-producing, of making and negotiating heritage values, often relies on more than verbal clues and sensory experiences that exceed narrative and representational canons. Co-curated memory mapping has proved to be a creative and inclusive way to gather strands of lived lives, past and present, which have solidified into multiple passions, dreams and convictions. Then, there is place. The starting point of my research was to bring to the fore places as leading agencies in the heritage experience alongside people and their deeds, actions and intentions. And the more I looked at the places, the more I discovered about the people who inhabited them, too, and vice versa. My main point was one of praise for this method for its simplicity. It is utterly easy, cheap and satisfying to enact with virtually any given community or social cluster of residents, commemorative group or place-based dwellers (Cross, 2012). The main point of adopting or indeed adapting this kind of methodology is that it can be done by virtually anyone, anywhere. Where country-specific restrictions on the production of geospatial or georeferenced data exist (e.g. Pakistan), alternative visualisation modes and productions can take place within the remit of creative, empowering community fieldwork.

What are the chief benefits of this experimental mapping technique? A map:

- Traces the history of meaningful locales as well as defines the actual and imagined topography of places
- Expresses a multitude of contemporary impressions, opinions and perceptions: it is inclusive and accretional
- Blends places, folk tales, material culture and memories in one straightforward, colourful and approachable medium 'at a glance'
- Can be printed out, photocopied and shared as an artefact in itself
- Can be edited and added to

In the context of the open-endedness of historical narratives, this type of map addresses and explores bodily interactions and offers a way to understand sensory engagement through a focus on the manner in which people move around, but also on their personal attachments and naming practices. I am reminded of the spontaneous perception of the row of rocks as a 'staircase' on the Monte Altare in Chapter 3 and the naming of the Airey Houses stairs as 'dancers' in Chapter 6.

What to do if things go wrong

From accusations of partiality to judgements on positionality and power relations, there are pitfalls embedded within any collaborative and participatory fieldwork and research. Further complications occur when working with memory in a social context where multiple versions of the past (and present) coexist: that is, in the vast majority of cases! Further, a focus and commitment to multivocality itself (e.g. Thompson, 1998; Dicks, 2010) does not always ensure the smooth running of social justice, and letting many voices speak at once does not automatically lead to unproblematic representation.

Public and private memory tend to occupy the same spaces of performance and perception, although the channels – and affects – through which these two strands of memory are animated differ (Crouch, 2010; Kidron, 2012; Macdonald, 2013). The tension between mainstream, public displays of heritage values and historical understandings that inform public and collective memorialisation (monuments, official celebrations and so on) are often at odds with grassroots, private understandings of the past (Cándida-Smith, 2003), be it the deep past as in Chapter 3 or the recent industrial past in Chapter 6. The relative informality of the collaborative mapping medium (as opposed to, for instance, the consultation of documents in an archive, which we cannot interrogate, or the interrogation of individual memory in oral history interviews) may unlock hitherto undisclosed information and versions of events, thus boosting plurality and open-endedness in the research process. And yet, as with any kind of memory and recollection work, there will be elements of contradiction and conflict, and a number of alternate versions challenging mainstream remembrance narrative(s) are bound to emerge (Portelli, 2003; Suleiman, 2008; Wertsch, 2008). In the case of the Italian civil war, where the veteran partisans have thus far effectively appropriated the mainstream public arena, we may ask whether multiple voices might be more likely to conflict with each other and create ingroups and outgroups in a mnemonic landscape where some experiences are more dominant than others.

In all memory work, collaboration itself entails challenges. For example, conflicting or irreconcilable memories may emerge. Correspondingly, there are intrinsic limitations to the community mapping method akin to the challenges occurring whenever we do fieldwork in which a number of people participate at different times. There may be disagreement as to what is relevant or not, and there will be a form of dissent over which (whose?) content ought to appear on the relevant visualisations. Some co-researchers' personal viewpoints and perspectives will likely be at odds with currently held views and perspectives, and the expression of these discordant voices may engender tension or inhibit sharing. Memories do not always align, and dissent can lead to non-collaboration, not forgetting that even the same person can recall things differently at different points in time and contradict themselves (Rouverol, 1999).

It may be good to honestly reflect on whether the visualisation and mapping methodology facilitates these conflicts. In the design and implementation of all phases, from fieldwork to editing of content, collaborative map-makers and activists establish a rapport. There are issues to consider as we facilitate these projects, namely issues of building rapport, access to resources, logistics and time restrictions. We need to allow for unforeseen cancellations, member withdrawals and co-researchers' refusal to allow the sharing of resources. This is not something that only happens when mapping processes are implemented, as unpredictability may affect fieldwork at some point or other in our careers and practice. The instability and unpredictability of carrying out research with others in the community – in the real world – however, should not discourage us from the otherwise positive, creative and inclusive process of making a map or visualising memory and heritage resources with others. There are strategies we can implement without compromising the integrity or the thoroughness of the process.

Making these maps has not always made for smooth sailing – if not in the field, in the aftermath, at the editing desk and community forum level. I have experienced logistical and theoretical, and often ethical, hurdles in the running of the process, processes which I explore in greater depth in the appendix. For instance, the range and extent of community co-researchers' and visitors' bodily engagements depend on their autobiographical materialities: that is, their individual stories, age, ability, previous experience and investment in any given locale or site. Some may be excluded from heritage spaces and memory sites due to the geographical and logistical nature or layout of locales incompatible with different abilities. The imagination, in that case, may still fill the gaps, and the sense of place felt towards a locale may still be meaningful to a person incapable of physically moving through the memoryscape. Their engagement may feel or look different, but it can be equally involving and satisfying.

Whether explored singularly or as a group, locations and places still do not make for easy 'processing'. Each experiment entails a continued negotiation of values, expectations and identities: thus, the editing content selection is closely linked to the necessity of maintaining a balanced representation and negotiation among groups. The first concern for myself and my co-researchers and fellow cartographers was the wish to foreground on the maps and experiments only those aspects and content that would make the community happy or at least satisfy them. And even in this endeavour we clashed about and against issues of individual remembrance, competing perspectives and, simply, the practical restriction to limit our expressions to one or more sheets of A3 paper (mostly, but not always, selected due to the format's portability). The stories told and enacted in fieldwork, far from being linear narratives constituting database-friendly patterns and themes, present us with a world of sensations, impressions and hardly quantifiable emotions such as fear, regret, nostalgia and anger. Further, rather than telling stories, map-makers tend to 'relive' and retrace their steps into their world of sensation, creating an experience which deeply involves the researcher(s).

What constitutes, then, a so called 'deep' map? Chapter 3 described the initial drive that gave life to the visualisations I made with Archaeoclub communities as a desire to expressly tell local stories that the professional archaeologists had ignored or banished from their outputs. So, memories, sensory experiences, impressions, alternative and playful imaginings of heritage sites and landscapes (remember Carlo's sleeping giant in Chapter 3?), all happily coexisted on these early maps along with chronological data and artefacts. The idea was to express non-conventional imaginaries of places with a past commingling with archaeologies of the present and to free longue durée landscapes from the restrictions of archaeological outputs. Everyone in the community with an interest in a site got involved, playing around with ideas and pictures and words and memories and proudly contributing to an artefact (the map) they knew would end up being examined in the lofty halls of a Russell Group university in the UK: there was much excitement about that possibility, as I recall.

Overall, the emphasis on tangible and intangible, presence and absence in the visualisations we have witnessed challenges the established conventions of quantitative fieldwork analysis. This kind of community-centred visualisation and mapping foregrounds and positively gathers together 'data' from a wider range of sources in novel and imaginative ways. This approach also questions the nature of data, opening it up to an array of imaginaries, hopes and dreams that have no tethering to visible forms or buildings as is common in traditional heritage and cultural-geographical outputs (see but Kiddey, 2017).

Moving forward: a call for action

To sharpen and reframe our understanding of memory, of living heritage, of recollections of industry or war or the way things have been, means engaging with present understandings in a hands-on, proactive way. Reframing and sharing stories means getting one's hands dirty with glue or ink or soil, navigating places, capturing sensations, shaping new memories in the everyday. Despite new efforts to engage communities and open up collaborative scholarship in the historical sciences, more work is needed to develop new ways of ensuring empowerment and representation of community values that exceed the quantitative or representational canon. We can do more. Scholars and fieldworkers wishing to involve communities to create co-produced outputs and narratives should be able to do so from the beginning of a research project (see also Cross, 2012). We should expand the range of communities that we work with, seeking to give a voice to, and empower, ever more marginalised or silenced groups. Professionally speaking, we should write co-production and collaborative fieldwork so that we enlist citizen science as a framework for knowledge-building, not as a footnote on methods (Allbrook and McGrath, 2015; Kennedy et al., 2016).

A further consideration on this practice: I can only speak (and write) from my own experience, and the insights and arguments I presented in this book serve as observations from my autobiographical standpoint. From doctoral student experiment to professional 'modus operandi', the mapping and visualisation experiments have been with me through the years and helped me think through making and showing data and stories. I am obviously partial in advocating this methodology, but not in order to feel clever – rather, because I have honestly found that it works. The collaborative fieldwork-and-mapping strategy is effective in engaging the participation and stimulating the curiosity of avocational members of the public in what 'heritage' and 'memory work' can accomplish. To perceive themselves as the architects and gatekeepers of knowledge leads to the public's sense that the workings of academic production and heritage policy have to do with them. The feeling that the field methods are centred around their experience, rather than milking them for information, may boost communities' sense of worth and reinforce their pride of place.

The process of mapping has not only bridged the gap of communication and interaction with the communities in which I found myself working, but it has also shaped my thinking and sharpened my scholarly intuitions. As a teaching tool, I love getting students enthused about this fun way of engaging communities in their research and fieldwork. All students of mine have thoroughly enjoyed getting involved with real-world things, textures, images and impressions. Co-researchers and learners alike have expressed an enjoyment in playing with data and information in a slightly different way. This is the point where I modestly suggest that readers with a role in education and training in the historical and social sciences and community-facing arts and humanities should think about recommending this method for their own students and trainees.

I attempt to sum up some of the findings of this technique and its application as follows:

- It is easy, cheap and quick to assemble and produce
- No advance knowledge of computer or image editing is necessary

- Groups of all ages and abilities can get involved
- It benefits and involves local communities: stakeholders who feel they belong there directly or indirectly literally put their own geographical and cultural values 'back on the map'
- It can be integrated, or form the foundation of, formal academic outputs and publications and student papers and projects – from summative report to doctoral dissertation

Figures

References

Abraham, N. and Torok, M. (1994). [Ecorce et le Noyau] *The shell and the kernel: Renewals of psychoanalysis*. Edited, translated and with an introduction by N. Rand. Chicago: University of Chicago Press.

Abu-Lughod, L. (2010). Return to half-ruins: Memory, postmemory, and living history in Palestine. *Journal of Comparative Poetics*, pp. 1–45.

Agnew, J. (1987). *Place and politics*. Boston, MA: Allen & Unwin.

Ahmad, Y. (2006). The scope and definitions of heritage: From tangible to intangible. *International Journal of Heritage Studies,* 12(3), pp. 292–300.

Ahmed, S. (2004). Affective economies. *Social Text,* 22, pp. 114–139.

Ahmed, S. (2010). Happy objects. In: M. Greg and G.J. Seigworth, eds., *The affect theory reader*. Durham, NC: Duke University Press, pp. 29–51.

Aksam, T. (2007). *A shameful act: The Armenian genocide and the question of Turkish responsibility*. New York: Holt.

Alaimo, S. (2008). Trans-corporeal feminisms and the ethical space of nature. In: S. Alaimo and S. Hekman, eds., *Material feminisms*. Bloomington: Indiana University Press, pp. 237–264.

Alcock, S. (2002). *Archaeologies of the Greek past: Landscape, monuments and memories*. Cambridge: Cambridge University Press.

Alcoff, L. (1991). The problem of speaking for others. *Cultural Critique,* 20, pp. 5–32.

Allbrook, M. and McGrath, A. (2015). Collaborative histories of the Willandra lakes. Deepening histories and the deep past. In: A. McGrath and M.A. Jebb, eds., *Long history, deep time: Deepening histories of place*. Canberra: ANU Press, pp. 41–252.

Alsayyad, N. (2001). *Hybrid urbanism: On the identity discourse and the built environment*. London: Routledge.

Anderson, B. (1983). *Imagined communities: Reflections on the origin and spread of nationalism*. London: Verso.

Anderson, B. (2009). Affective atmospheres. *Emotion, Space and Society,* 2(2), pp. 77–81.

Anderson, B. and Harrison, P. (eds.) (2010a). *Taking-place: Non-representational theories and geography*. Farnham: Ashgate.

Anderson, B. and Harrison, P. (2010b). The promise of non-representational theories. In: B. Anderson and P. Harrison, eds., *Taking – place: Non representational theories and geography*. Farnham: Ashgate, pp. 1–36.

Anderson, J. (2004). Talking whilst walking: A geographical archaeology of knowledge. *Area,* 36(3), pp. 254–261.

Andrews, G., Kearns, R., Kontos, P. and Wilson, V. (2006). Their finest hour: Older people, oral histories, and the historical geography of social life. *Social & Cultural Geography,* 7(2), pp. 153–177.

Andriotis, K. (2008). Sacred site experience: A phenomenological study. *Annals of Tourism Research*, 36(1), pp. 64–84.

Annis, S. (1986). The museum as a staging ground for symbolic action. *Museum International*, 38(3), pp. 168–171.

Antze, P. and Lambek, M. (1996). Introduction: Forecasting memory. In: P. Antze and M. Lambek, eds., *Tense past: Cultural essays in trauma and memory*. London: Routledge, pp. x–xxxviii.

Arnosti, G. (1993). Reperti votivi e santuari dei Paleoveneti nell'alto Cenedese. *Il Flaminio – Rivista di Studi Storico-Archeologici della Civiltà Montana delle Prealpi Trevigiane*, 6, pp. 55–82.

Ashworth, G.J. (1996). *Dissonant heritage: The management of the past as a resource in conflict*. New York: Belhaven Press.

Atalay, S. (2006). Indigenous archaeology as decolonizing practice. *The American Indian Quarterly*, 30(3), pp. 280–310. doi: 10.1353/aiq.2006.0015.

Atalay, S. (2007). Global application of indigenous archaeology: Community based participatory research in Turkey. *Archaeologies Arch*, 3(9), pp. 249–270.

Atalay, S. (2012). *Community-based archaeology: Research with, by, and for indigenous and local communities*. Berkeley: University of California Press.

Atia, N. (2010). A relic of its own past: Mesopotamia and the British imagination. *Memory Studies*, 3, pp. 232–241.

Atkinson, D. (2007). Kitsch geographies and the everyday spaces of social memory. *Environment and Planning A*, 39, pp. 421–440.

Babadzan, A. (2000). Anthropology, nationalism and 'the invention of tradition'. *Anthropological Forum*, 10(2), pp. 131–155.

Bagnall, G. (2003). Performance and performativity at heritage sites. *Museum and Society*, 1(2), pp. 87–103.

Ballone, A. (2007). *La Resistenza*. In: M. Isnenghi (2010), ed., *I luoghi della memoria. Strutture ed eventi dell'Italia Unita*. Rome: Laterza. Available at: https://docs.google.com/viewer?a=v&pid=sites&srcid=ZGVmYXVsdGRvbWFpbnxtYXN0ZXJlc3R1ZGl vc2l0YWxpYW5vvc3xneDo3YzdiNjRjMDU1MWQ5MzBk&pli=1.

Ball-Rokeach, S.J., Kim, Y.-C. and Matei, S. (2001). Storytelling neighbourhood: Paths to belonging in diverse urban environments. *Communication Research*, 28(4), pp. 392–428.

Barad, K. (2007). *Meeting the universe half way*. Durham, NC: Duke University Press.

Barad, K. (2008). Posthumanist performativity: Towards and understanding of how matter comes to matter. In: S. Alaimo and S. Hekman, eds., *Material feminisms*. Bloomington: Indiana University Press, pp. 120–156.

Barbour, L. (2016). Rodenbach, grandfather and me: Echoes and premonitions of the First World War. *Emotion, Space and Society*, 19, pp. 94–102.

Barry, M. (2014). 'Please do touch: Discourses on aesthetic interactivity in the exhibition space'. *Participations: International Journal of Audience Research*, 11(1), pp. 216–235.

Battaglia, R. (1953). *Storia della Resistenza Italiana*. Turin: Einaudi.

Beauregard, R. (2013). The neglected places of practice. *Planning Theory & Practice*, 14, pp. 8–19.

Behan, T. (2009). *The Italian resistance. Fascists, guerrillas and the allies*. London: Pluto Press.

Belcher, O., Martin, E., Secor, A., Simon, S. and Wilson, T. (2008). Everywhere and nowhere: The exception and the topological challenge to geography. *Antipode*, 40(4), pp. 499–503.

Bell, D. (2006). Introduction. In: D. Bell, ed., *Memory, trauma and world politics. Reflections on the relationship between past and present*. Basingstoke: Palgrave, pp. 1–29.

Bell, M. M. (1997). The ghosts of place. *Theory and Society*, 26, pp. 813–836.

Bender, B. (1993b). Introduction. In: B. Bender, ed., *Landscape: Politics and perspectives*. Oxford: Berg, pp. 1–18.

Bender, B. (2002). Time and landscape. *Current Anthropology,* 43(S4), Special Issue Repertoires of Timekeeping in Anthropology, pp. 103–112.

Bender, B. (2006). Place and landscape. In: C. Tilley, et al., ed., *Handbook of material culture*. London: Sage, pp. 303–314.

Bennett, J. (2010). *Vibrant matter: A political ecology of things*. Durham, NC: Duke University Press.

Benyon, H., Hudson, R. and Sadler, D. (1986). Nationalised industry policies and the destruction of communities: Some evidence from North East England. *Capital and Class,* 10(2), pp. 27–57.

Berliner, D. (2005). The abuses of memory: Reflections on the memory boom in anthropology. *Anthropological Quarterly,* 78(1), pp. 197–211.

Bessel, R. (ed.) (1996). *Fascist Italy and Nazi Germany*. Cambridge: Cambridge University Press.

Black, M. (2003). Expellees tell tales. *History and Memory,* 25(1), pp. 77–110.

Blake, K. S. (2005). Mountain symbolism and geographical imaginations. *Cultural Geographies,* 12(4), pp. 527–531.

Blum, V.L. and Secor, A.J. (2014). Mapping trauma: Topography to topology. In: P. Kingsbury and S. Pile, eds., *Psychoanalytic geographies*. Farnham: Ashgate.

Blyth, A. (2015). School As Space, n.p. Online at https://alastairblyth.files.wordpress.com/2015/03/blyth-school-as-space.pdf Accessed 16 July 2019.

Bocca, G. (2008 [1945]). *Partigiani della Montagna. Vita delle Divisioni Giustizia e Libertà del Cuneese*. Milan: Feltrinelli.

Bondi, L. (2005). Making connections and thinking through emotions: Between geography and psychotherapy. *Transactions of the Institute of British Geographers,* 30(4), pp. 433–448.

Bondi, L., Davidson, J. and Smith, M. (2005). Introduction: Geography's 'emotional turn'. In J. Davidson, M. Smith and L. Bondi, eds., *Emotional geographies*. Aldershot: Ashgate, pp. 1–16.

Bonnett, A. (2010). *Left in the past: Radicalism and the politics of nostalgia*. New York: Continuum.

Boss, P. (2000). *Ambiguous loss: Learning to live with unresolved grief*. Harvard: Harvard University Press.

Boss, P. (2010). The trauma and complicated grief of ambiguous loss. *Pastoral Psychology,* 59(2), pp. 137–145.

Bourdieu, P. (1977). *Outline of a theory of practice*. Cambridge: Polity Press.

Brescacin, P.P. (2012). *Quel Sangue che Abbiamo Dimenticato. Volume I*. Vittorio Veneto: Tipse.

Brescacin, P.P. (2014). *Quel Sangue che Abbiamo Dimenticato. Volume II*. Vittorio Veneto: Tipse.

Brown, V. (2015). Mapping a slave revolt: Visualizing spatial history through the archives of slavery. *Social Text,* 33(4), pp. 134–141.

Buchli, V. and Lucas, G. (2001). The absent present: Archaeologies of the contemporary past. In: V. Buchli and G. Lucas, eds., *Archaeologies of the contemporary past*. London: Routledge, pp. 3–18.

Burkitt, I. (2012). Emotional reflexivity: Feeling, emotion and imagination in reflexive dialogues. *Sociology,* 46(3), pp. 458–472.

Butler, B. (2009). 'Othering' the archive – from exile to inclusion and heritage dignity: The case of the Palestinian archival memory. *Archival Science,* 9, pp. 57–69.

Butler, J. (1993). *Bodies that matter: On the discursive limits of 'sex'*. London and New York: Routledge.

Butler, J. (1997). *Excitable speech: A politics of the performative*. London: Routledge.

Byrd, J. (2011). *The transit of empire: Indigenous critiques of colonialism*. Minneapolis: University of Minnesota Press.

Cahill, C., Sultana, F. and Pain, R. (2007). Editorial. *ACME: An International E-Journal for Critical Geographies,* 6(3), pp. 304–318.

Cairns, S. and Birchall, D. (2013). Curating the digital world: Past preconceptions, present problems, possible futures. In: N. Proctor and R. Cherry, eds., *Museums and the web.* Silver Spring, MD: Museums and the Web. Published Feb. 6.

Callinicos, A. (2004). *Making history: Agency, structure, and change in social theory*. Leiden: Brill.

Calvino, I. (1964). *Il Sentiero dei Nidi di Ragno*. Turin: Einaudi.

Cameron, E. (2012). New geographies of story and storytelling. *Progress in Human Geography,* 36(5), pp. 573–592.

Cándida-Smith, R. (2002 [2003]). Introduction: Performing the archive. In: R. Cándida-Smith, ed., *Art and the performance of memory: Sounds and gestures of recollection.* New York: Routledge, pp. 1–12.

Canuto, M. and Yaeger, J. (eds.) (2000). *The archaeology of communities: A new world perspective*. London: Routledge.

Cappelletto, F. (2003). Long-term memory of extreme events: From autobiography to history. *Journal of the Royal Anthropological Institute,* 9, pp. 241–260.

Cappelletto, F. (2005). Public memories and personal stories: Recalling the Nazi-Fascist massacres. In: F. Cappelletto, ed., *Memory and World War II: An ethnographic approach.* Berg: London, pp. 1–38.

Cashman, R. (2006). Critical nostalgia and material culture in Northern Ireland. *Journal of American Folklore,* 119(427), pp. 137–160.

Cashman, R. (2008). *Storytelling on the Northern Irish border: Characters and community*. Bloomington: Indiana University Press.

Castañeda, Q.E. (2008). The 'ethnographic turn' in archaeology: Research positioning and reflexivity in ethnographic archaeologies. In: Q.E. Castañeda and C.N. Matthews, eds., *Ethnographic archaeologies: Reflections on stakeholders and archaeological practice.* Lanham, MD: Altamira Press, pp. 25–62.

Castañeda, Q.E. and Matthews, C.N. (2008b). Introduction: Ethnography and the social construction of archaeology. In: Q.E. Castañeda and C.N. Matthews, eds., *Ethnographic archaeologies: Reflections on stakeholders and archaeological practice.* Lanham, MD: Altamira Press, pp. 2–24.

Chadwick, A. and Gibson, C. (eds.) (2013a). *Memory, myth and long-term landscape.* Oxford: Oxbow Books.

Chadwick, A. and Gibson, C. (2013b). Do you remember the first time? In: A. Chadwick and C. Gibson, eds., *Memory, myth and long-term landscape.* Oxford: Oxbow Books, pp. 1–31.

Clark, M. (1984). *Modern Italy, 1871–1982*. London: Longman.

Classen, C. and Howes, D. (2006). The sensescape of the museum: Western sensibilities and indigenous artifacts. In: E. Edwards, C. Gosden and R. Phillips, eds., *Sensible objects: Colonialism, museums and material culture*. Oxford: Berg, pp. 199–220.

Cole, T. (2015). (Re)Placing the past: Spatial strategies of retelling difficult stories. *Oral History,* 42(1), pp. 30–49.

Connerton, P. (1989). *How societies remember*. Cambridge: Cambridge University Press.

Connor, M.C. (2005). *The invention of terra nullius: Historical and legal fictions on the foundation of Australia*. Sydney: Macleay Press.

Conway, M. (1996). Autobiographical memory. In: E.L. Bjork and B.L. Bjork, eds., *Memory. handbook of perception and cognition, volume 2*. San Diego: Academic Press, pp. 165–196.

Cooke, P. (2011). *The legacy of the Italian resistance*. Basingstoke: Palgrave MacMillan.

Crampton, J. (2001). Maps as social constructions: Power, communication and visualization. *Progress in Human Geography*, 25, pp. 235–252.

Crampton, J. (2010). *Mapping: A critical introduction to cartography and GIS*. Oxford: Blackwell.

Cresswell, T. (2004). *Place: A short introduction*. Oxford: Blackwell.

Cross, S. (2012). *Look, listen and learn: Back in Kilkenny: Hurling and community heritage mapping*. Available at: http://susancrosstelltale.com/2012/09/15/look-listen-and-learn-community-heritage-mapping/ [Accessed 1 July 2018].

Crouch, D. (2010). The perpetual performance and emergence of heritage. In: S. Watson and E. Waterton, eds., *Culture, heritage and representation: Perspectives on visuality and the past*. Farnham: Ashgate, pp. 57–71.

Crouch, D. (2015). Affect, heritage, feeling. In: E. Waterton and S. Watson, eds., *The Palgrave handbook of contemporary heritage research*. Basingstoke: Palgrave Macmillan, pp. 177–190.

Cruikshank, J. (2006). *Do glaciers listen? Local knowledge, colonial encounters, and social imagination*. Vancouver: UBC Press.

Cubitt, G. (2007). *History and memory*. Manchester: Manchester University Press.

Curti, G.H. (2008). From a wall of bodies to a body of walls: Politics of affect – politics of memory – politics of war. *Emotion, Space and Society*, 1, pp. 106–118.

de Certeau, M. (1984). *The practice of everyday life*. Berkeley: University of California Press.

De Felice, R. (1998). *Mussolini l'Alleato: la Guerra Civile 1943–45*. Turin: Edizioni Einaudi.

De Nardi, S. (2014a). Senses of place, senses of the past: Making experiential maps as part of community heritage fieldwork. *Journal of Community Archaeology and Heritage*, 1(1), pp. 5–23.

De Nardi, S. (2014b). 'No one had asked me about that before': A focus on the body and 'other' resistance experiences in Italian World War Two storytelling. *Oral History*, 43(1), pp. 5–23.

De Nardi, S. (2014c). An embodied approach to Second World War storytelling mementoes: Probing beyond the archival into the corporeality of memories of the resistance. *Journal of Material Culture*, 19(4), pp. 443–464.

De Nardi, S. (2015). When family and research clash: The role of autobiographical emotion in the production of stories of the Italian civil war, 1943–1945. *Emotion Space and Society*, 17, pp. 22–29.

De Nardi, S. (2016). *The poetics of conflict experience: Materiality and embodiment in Second World War Italy*. London: Routledge.

De Nardi, S. (2019). Everyday heritage activism in Swat valley: Ethnographic reflections on a politics of hope. *Heritage and Society*. doi: 10.1080/2159032X.2018.1556831; https://doi.org/10.1080/2159032X.2018.1556831

Dedrick, M. (2018). Photovoice as a method for the development of collaborative archaeological practice. *Journal of Community Archaeology and Heritage*, 5(1). https://doi.org/10.1080/20518196.2018.1442659

Del Casino, V. and Hanna, S. (2006). Beyond the 'binaries': A methodological intervention for interrogating maps as representational practices. *ACME*, 4(1), pp. 34–56.

DeLanda, M. (2006). *A new philosophy of society: Assemblage theory and social complexity*. London: Bloomsbury/Continuum Press.

Deleuze, G. (1994). *Difference and repetition*. New York: Columbia University Press.

Deleuze, G. and Guattari, F. (2004). *A thousand plateaus*. Translation by B. Massumi. Minneapolis: University of Minnesota Press.

Della Libera, A. (1988). *Sulle Montagne per la Libertà*. Vittorio Veneto: Tipse.

DeLyser, D. (1999). Authenticity on the ground: Engaging the past in a California ghost town. *Annals of the Association of American Geographers*, 89(4), pp. 602–632.

DeSilvey, C. (2006). Observed decay: Telling stories with mutable things. *Journal of Material Culture*, 11(3), pp. 318–338.

DeSilvey, C. (2012). Making sense of transience: An anticipatory history. *Cultural Geographies*, 19(1), pp. 31–54.

Di Scala, S. (1999). Resistance mythology. *Journal of Modern Italian Studies*, 4(1), pp. 67–72.

Dicks, B. (2010). *Heritage, place and community*. Cardiff: University of Wales Press.

Dicks, B. (2015). Heritage and social class. In E. Waterton and S. Watson, S. eds., *The Palgrave handbook of contemporary heritage research*. Basingstoke: Palgrave, pp. 366–438.

Dicks, B. (2016). The habitus of heritage: A discussion of Bourdieu's ideas for visitor studies in heritage and museums. *Museum and Society*, 14(1), pp. 52–64.

Domanska, E. (2006). The material presence of the past. *History and Theory*, 45(3), pp. 337–348.

Drozdzewski, D. (2012). Knowing (or not) about Katyń: The silencing and surfacing of public memory. *Space and Polity*, 16, pp. 303–319.

Drozdzewski, D. (2015). Retrospective reflexivity: The residual and subliminal repercussions of researching war. *Emotion, Space and Society*. Available online 24 Apr. 2015. doi: 10.1016/j.emospa.2015.03.004.

Drozdzewski, D. and Birdsall, C. (2019). Advancing memory methods. In: D. Drozdzewski and C. Birdsall, eds., *Doing memory research: New methods and approaches*. London: Palgrave, pp. 1–20.

Dudley, S. (ed.) (2010). *Museum materialities: Objects, engagements, interpretations*. New York: Routledge.

Edwards, E. (2010). Photographs and history: emotion and materiality. In S. Dudley, ed., *Museum materialities*. London: Routledge, pp. 21–38.

Fenster, T. and Misgav, C. (2014). Memory and place in participatory planning. *Planning Theory & Practice*, 15(3), pp. 349–369.

Fentress, J. and Wickham, C. (1992). *Social memory*. Oxford: Blackwell.

Ferris, N. and Welch, J. (2014). Beyond archaeological agendas: In the service of a sustainable archaeology. In S. Atalay, L.R. Clauss, R. H. McGuire and J. Welch, eds., *Transforming Archaeology: Activist practices and prospects*. Walnut Creek: Left Coast Press, pp. 215–238.

Field, S. (2008). Imagining communities: Memory, loss, and resilience in post-apartheid Cape Town. In: P. Hamilton and L. Shopes, eds., *Oral history and public memories*. Philadelphia: Temple University Press.

Field, S. (2014). Loose bits of shrapnel: War stories, photographs, and the peculiarities of postmemory. *Oral History Review*, 41(1), pp. 108–131.

Fieldhouse, E. and Hollywood, E. (1999). Life after mining: Hidden unemployment and changing patterns of economic activity amongst miners in England and Wales, 1981–1991. *Work, Employment and Society*, 13(3), pp. 483–502.

Flood, F.B. (2002). Between cult and culture: Bamiyan, Islamic iconoclasm, and the museum. *The Art Bulletin*, 84(4), pp. 641–659.

Focardi, F. (2013). *Il Cattivo Tedesco e il Bravo italiano*. Rome: Laterza.

Fowler, C. (2013). *The emergent past: A relational realist archaeology of early bronze age mortuary practices*. Oxford: Oxford University Press.

Fox, N.J. and Alldred, P. (2018). The materiality of memory: Affects, remembering and food decisions. *Cultural Sociology*, pp. 1–17.

Gallaher, C. (2016). Placing the militia occupation of the Malheur national wildlife refuge in Harney county, Oregon. *ACME*, 15(2), pp. 293–308.

Gambacurta, G. and Gorini, G. (2005). La stipe votiva del Monte Altare. In: G. Gorini and A. Mastrocinque, eds., *Stipi Votive delle Venezie*. Rome: Giorgio Bretschneider, pp. 150–210.

Gamboni, D. (2001). World heritage: Shield of target? *Getty Conservation Institute Newsletter*, 16, pp. 5–11.

Gibson, K. (2001). Regional subjection and becoming. *Environment and Planning D*, 19(6), pp. 639–667.

Ginsborg, P. (1990). *A history of contemporary Italy. Society and politics 1943–1988*. London: Penguin.

Gosden, C. (2005). What do objects want? *Archaeological Method and Theory*, 12(3), pp. 193–201.

Graaffland, A. (1999). Of rhizomes, trees, and IJ-Oevers, Amsterdam. *Assemblage*, 38, pp. 28–41.

Graham, H.C. (2016). The 'co' in co-production: Museums, community participation and Science and Technology studies. *Science Museum Group Journal*. doi:10.15180/160502

Gregory, D. (1994). *Geographical imaginations*. Oxford: Blackwell.

Gregory, K. and Witcomb, A. (2007). Beyond nostalgia: The role of affect in generating historical understanding at heritage sites. In: S.J. Knell, S. Macleod and S. Watson, eds., *Museum revolutions: How museums change and are changed*. London: Routledge, pp. 263–275.

Grey, C. (2008). Instrumental policies: Causes, consequences, museums and galleries. *Cultural Trends*, 17(4), pp. 209–222.

Grosz, E. (2008). Darwin and feminism: Preliminary investigations for a possible alliance. In: S. Alaimo and S. Hekman, eds., *Material feminisms*. Bloomington: Indiana University Press, pp. 23–51.

Halbwachs, M. (1992 [1925]). *Les Cadres Sociaux de la Memoire*. Paris: Gallimard.

Halbwachs, M. (1997 [1950 Posthumous]). *La Mémoire collective*. Paris: Albin Michel.

Hall, S. (ed.) (1997). *Cultural representations and signifying practices*. London: Sage.

Hall, S. (2005). Whose heritage? Un-settling "the heritage", re-imagining the post-nation. In: J. Littler and R. Naidoo, eds., *The politics of heritage: The legacies of "race"*. London: Routledge, pp. 23–35.

Hamilton, S., Whitehouse, R., Brown, K., Herring, E. and Seager-Thomas, M. (2006). Phenomenology in practice: Towards a methodology for a 'subjective' approach. *European Journal of Archaeology*, 9(1), pp. 31–71.

Hancock, N. (2010). Virginia Woolf's glasses: material encounters in the literary/artistic house museum. In S. Dudley, ed., *Museum materialities. Objects, engagements, interpretations*. London: Routledge, pp. 114–127.

Hardy, K. (2012). Dissonant emotions, divergent outcomes: Constructing space for emotional methodologies in development. *Emotion, Space and Society*, 5, pp. 113–121.

Harley, B. (1989). Deconstructing the map. *Cartographica*, 26(2), pp. 1–20.

Harley, B. (1992). Deconstructing the map. In T. Barnes and J. Duncan. *Writing worlds: Discourse, text and metaphor in the representation of landscape*. London: Routledge, pp. 231–247.

Harley, J.B. and Laxton, P. (eds.) (2001). *The new nature of maps: Essays in the history of cartography*. Baltimore: John Hopkins University Press.

Harris, O. (2014). Re-assembling communities. *Journal of Archaeological Method and Theory*, 21, pp. 76–97.

Haslam, N., Bain, P., Douge, L., Lee, M. and Bastian, B. (2005). More human than you: Attributing humanness to self and others. *Journal of Personality and Social Psychology*, 89(6), pp. 937–950.

Hekman, S. (2008). Constructing the ballast: An ontology for feminism. In: S. Alaimo and S. Hekman, eds., *Material feminisms*. Indianapolis: University of Indiana Press, pp. 85–119.

Heller, A. (2001). A tentative answer to the question: Has civil society cultural memory? *Social Research*, 68(4), pp. 103–142.

Henare, A.J., Holbraad, M. and Wastell, S. (2007). Introduction. In A. Henare, M. Holbraad and S. Wastell, eds., *Thinking through things: Theorising artefacts ethnographically*. London: Routledge, pp. 1–31.

Heritage Lottery Fund, n.d. https://www.heritagefund.org.uk/our-work/community-heritage

Hillman, D. and Mazzio, C. (1997). *The body in parts: Fantasies of corporeality in early modern Europe*. New York: Routledge.

Hirsch, M. (1996). Past lives: Postmemories in exile. *Poetics Today*, 17(4), pp. 659–686.

Hirsch, M. (2008). The generation of postmemory. *Poetics Today*, 29, pp. 103–128.

Hobsbawm, E. and Ranger, T. (eds.) (1993). *The invention of tradition*. Cambridge: Cambridge University Press.

Hodgkin, K. and Radstone, S. (2003). Introduction: Contested pasts. In: K. Hodgkin and S. Radstone, eds., *Contested pasts: The politics of memory*. Routledge: London, pp. 1–22.

Horning, A. (2013). Exerting influence? Responsibility and the public role of archaeology in divided societies. *Archaeological Dialogues*, 20(1), pp. 19–29.

Horning, A. and Breen, C. (2017). In the aftermath of violence: Heritage and conflict transformation in Northern Ireland. In: P. Newson and R. Young, eds., *Post-conflict archaeology and cultural heritage*. London: Routledge, pp. 177–194.

Hrobat, K. (2007). Use of oral tradition in archaeology: The case of Ajdovščina above Rodik, Slovenia. *European Journal of Archaeology*, 10(1), pp. 31–56.

Hrobat Virloget, K., Poljak Istenič, S., Čebron Lipovec, N. and Habinc, M. (2016). Abandoned spaces, mute memories: On marginalized inhabitants in the urban centres of Slovenia. *Proceedings of the SANU Ethnographic Institute Гласник Етнографског института САНУ*, 64(1), pp. 77–90.

Iacono, F. and Këlliçi, K.L. (2015). Of pyramids and dictators: Memory, work and the significance of communist heritage in post-socialist Albania. *AP: Online Journal in Public Archaeology*, 5, pp. 97–122.

Ingold, T. (1993). The temporality of the landscape. *World Archaeology*, 25(2), pp. 152–174.

Ingold, T. (2000). *The perception of the environment*. Oxford: Berg.

Ingold, T. (2011). *Being alive. Essays on movement, knowledge and description*. London: Routledge.

Ingold, T. (2007). Materials against materiality. *Archaeological Dialogues*, 14(1), pp. 1–16.

Jarrett, K. (2013). Telling tales? Myth, memory and Crickley Hill. In: A. Chadwick and C. Gibson, eds., *Memory, myth and long-term landscape*. Oxford: Oxbow Books, pp. 189–208.

Johnson, M. (2006). *Ideas of landscape*. London: Wiley.

Jones, A.M. (2007). *Memory and material culture*. Cambridge: Cambridge University Press.

Jones, O. and Garde-Jansen, J. (eds.) (2012). *Geography and memory: Explorations in identity, place and becoming*. Basingstoke: Palgrave Macmillan.

Kavanagh, G. (2000). *Dream spaces: Memory and the museum*. London: Leicester University Press.

Kearney, R. (1998). *Poetics of imagining*. New York: Fordham University Press.

Keller, S. (2003). *Community. Pursuing the dream, living in the reality*. Princeton: Princeton University Press.

Kennedy, R., Leane, J., McGrath, A., et al. (2016). Roundtable: Message from Mungo and the scales of memory. *Australian Humanities Review,* 59, pp. 247–258.

Kenway, J. and Fahey, J. (2009). Imagining research otherwise. In: J. Kenway and J. Fahey, eds., *Globalizing the Research Imagination*. New York: Routledge. pp. 1–40.

Kiddey, R. (2014). Punks and drunks: Counter mapping homeless heritage. In: J. Schofield, ed., *Who needs experts? Counter-mapping cultural heritage*. London: Routledge, pp. 165–180.

Kiddey, R. (2017). *Homeless heritage: Collaborative social archaeology as therapeutic practice*. Oxford: Oxford University Press.

Kiddey, R. and Schofield, J. (2011). Embrace the margins: Adventures in archaeology and homelessness. *Public Archaeology,* 10, pp. 4–22.

Kidron, C. (2012). Breaching the wall of traumatic silence: Holocaust survivor and descendant person – object relations and the material transmission of the genocidal past. *Journal of Material Culture,* 17(1), pp. 3–21.

Kirk, T. (1992). Space, subjectivity, power and hegemony: Megaliths and long mounds in earlier neolithic Brittany. In: C. Tilley, ed., *Interpretative archaeology*. Oxford: Berg, pp. 181–224.

Klinkhammer, L. (1996). *Stragi Naziste in Italia: 1944–1945*. Rome: Donzelli Editore.

Knapp, B. and Ashmore, W. (1999). Archaeological landscapes: Constructed, conceptualized, ideational. In: W. Ashmore and B. Knapp, eds., *Archaeologies of landscape*. Oxford: Blackwell, pp. 1–30.

Kohon, G. (2016). *Reflections on the aesthetic experience, psychoanalysis and the uncanny*. London: Routledge.

Korsmeyer, C. and Sutton, D. (2011). The sensory experience of food. *Food, Culture & Society,* 14(4), pp. 461–475.

Koshar, R. (2000). *From monuments to traces: Artifacts of German memory 1870–1990*. Berkeley: University of California Press.

Koskinen-Koivisto, E. (2016). Reminder of dark heritage of human kind: Experiences of Finnish cemetery tourists of visiting the Norvajärvi German cemetery. *Thanatos,* 5(1), pp. 23–41.

Kvale, S. and Brinkmann, S. (2009). *InterViews: Learning the craft of qualitative research interviewing* (2nd ed.). Thousand Oaks, CA: Sage Publications.

Kyvik, G. (2004). Prehistoric material culture, presenting, commemorating, politicising. In F. Fahlander and T. Oestigaard, eds., *Material culture and other things. Post–disciplinary studies in the 21st century*. Lindome: Bricoleur Press, pp. 87–102.

LaCapra, D. (2001). *Writing history, writing trauma*. Baltimore: John Hopkins University Press.

Langer, M. (1989). *Merleau – Ponty's phenomenology of perception: A guide and commentary*. Basingstoke: Macmillan.

Larkin, C. (2012). *Memory and conflict in Lebanon: Remembering and forgetting the past*. London: Routledge.

Latham, A. and McCormack, D. (2009). Thinking with images in non-representational cities: Vignettes from Berlin. *Area,* 41, pp. 252–262.

Latour, B. (2005). *Reassembling the social: An introduction to actor-network-theory.* Oxford: Oxford University Press.

Layolo, L. (1998). La guerra civile. In: V. Pianca, ed., *Geografia della Resistenza.* Vittorio Veneto: Tipse, pp. 80–84.

Legg, S. (2005). Contesting and surviving memory: Space, nation, and nostalgia in Les Lieux de Mémoire. *Environment and Planning D: Society and Space,* 23, pp. 481–504.

Levinas, E. (1998 [1991]). *Entre Nous: Thinking – of – the – other.* Translated by M.B. Smith and B. Harshaw. New York: Columbia University Press.

Leyens, J. (2009). Retrospective and prospective thoughts about infrahumanization. *Group Processes & Intergroup Relations,* 12(6), pp. 807–817.

Leyens, J., Paladino, P., Rodriguez-Torres, R., Vaes, J., Demoulin, S., Rodriguez-Perez, A. and Gaunt, R. (2000). The emotional side of prejudice: The attribution of secondary emotions to ingroups and outgroups. *Personality and Social Psychology Review,* 4(2), pp. 186–197.

Lieberman, S. (2015). Missing. *Emotion Space and Society,* 19, pp. 87–93.

Light, D. and Young, C. (2016). Multiple and contested geographies of memory: Remembering the 1989 Romanian 'revolution'. In: D. Drozdzewski, S. De Nardi and E. Waterton, eds., *Memory, place and identity: Commemoration and remembrance of war and conflict.* London: Routledge.

Lippard. L. (2000). *The lure of the local: Sense of place in a multicentered society.* New York: New Press.

Logan, W. (2012). Cultural diversity, cultural heritage and human rights: Toward heritage management as human rights-based cultural practice. *International Journal of Heritage Studies,* 18(3), pp. 231–244.

Logan, W. and Reeves, K. (2009). Introduction: Remembering places of pain and shame. In: W. Logan and K. Reeves, eds., *Places of pain and shame: Dealing with difficult heritage.* Abington: Routledge, pp. 1–14.

Louise, D. (2016). *A Q&A with . . . Nico Vascellari, artist and punk musician.* Available at: www.a-n.co.uk/news/a-qa-with-nico-vascellari-artist-and-punk-musician [Accessed 31 August. 2019].

Lowenthal, D. (1985). *The past is a foreign country.* Cambridge: Cambridge University Press.

Lucero, J. A. (2008). *Struggles of voice: The politics of indigenous representation in the Andes.* Pittsburgh: University of Pittsburgh Press.

Lyndall, R., Debenham, J., Brown, M. and Pascoe, W. (2017). *Colonial frontier massacres in eastern Australia 1788–1872.* University of Newcastle. Available at: https://c21ch. newcastle.edu.au/colonialmassacres/findings.php [Accessed 19 June 2017].

Maantay, J.A. (2013). The collapse of place: Derelict land, deprivation, and health inequality in Glasgow, Scotland. *Cities and the Environment,* 6(1). Article 10. http://digitalcom mons.lmu.edu/cate/vol6/iss1/10 [Accessed 30 Mar. 2019].

Macdonald, S. (1997). A people's story: Heritage, identity and authority. In: C. Rojek and J. Urry, eds., *Touring cultures: Transformations of travel and theory.* London: Routledge, pp. xx–xxx.

Macdonald, S. (2009). *Difficult heritage: Negotiating the Nazi past in Nuremberg and beyond.* London: Routledge.

Macdonald, S. (2013). *Memorylands: Heritage and identity in Europe today.* London: Routledge.

MacKian, S. (2004). Mapping reflexive communities: Visualising the geographies of emotion. *Social and Cultural Geography,* 5(4), pp. 615–631.

Mady, C. (2018). Public space activism in unstable contexts: Emancipation from Beirut's postmemory. In: S. Kierbein and T. Viderman, eds., *Public space unbound: Urban emancipation and the post-political condition*. London: Routledge.

Malpas, J. (1999). *Place and experience: A philosophical topography*. Cambridge: Cambridge University Press.

Mammone, A. (2006). A daily revision of the past: Fascism, anti-Fascism, and memory in contemporary Italy. *Modern Italy*, 11, pp. 211–226.

Margalit, G. (2010). *Guilt, suffering, and memory: Germany remembers its dead of World War Two*. Translated by H. Watzmann. Bloomington: Indiana University Press.

Massey, D. (1995). Places and their pasts. *History Workshop Journal*, 39, pp. 182–192.

Matthews, C. (2019). Ethnographic archaeology, routine archaeologies, and social justice research. *Journal of Community Archaeology and Heritage*. https://doi.org/10.1080/20518196.2019.1600234

Matthews, C. and McDavid, C. (2012). Community archaeology. In: N. Silberman, ed., *The Oxford companion to archaeology*, vol. 1. Oxford: Oxford University Press, pp. 336–340.

May, S., Orange, H. and Penrose, S. (2012). *The good, the bad and the unbuilt: Handling the heritage of the recent past*. Studies in Contemporary and Historical Archaeology. BAR Series 7. Oxford: Archaeopress.

McAtackney, L. (2014). *An archaeology of the troubles: The dark heritage of long kesh/maze prison*. Oxford: Oxford University Press.

McAtackney, L. (2015). Memorials and marching: Archaeological insights into segregation in contemporary Northern Ireland. *Historical Archaeology*, 49(3), pp. 110–125.

McAtackney, L. (2016). Re-remembering the troubles: Community memorials, memory and identity in post-conflict Northern Ireland. In: E. Epinoux and F. Healy, eds., *Post-Celtic tiger Ireland: Exploring new cultural spaces*. Cambridge: Cambridge Scholars Publishing, pp. 42–64.

McAtackney, L. (2018). Where are all the women? Public memory, gender and memorialisation in contemporary Belfast. In: E. Crooke, ed., *Post-conflict heritage*. London: Palgrave Macmillan, PAGES.

McCarthy, C. (2017). Incidental heritage: Difficult intangible heritages as collateral damage. *International Journal of Heritage Studies*, 23(1), pp. 52–64.

McCormack, D. (2003). An event of geographical ethics in spaces of affect. *Transactions of the Institute of British Geographers*, 28, pp. 488–507.

McCormack, D. (2010). Remotely sensing affective afterlives: The spectral geographies of material remains. *Annals of the Association of American Geographers*, 100, pp. 640–654.

McDavid, C. (2002). Archaeologies that hurt; descendents that matter: A pragmatic approach to collaboration in the public interpretation of African-American archaeology. *World Archaeology*, 34(2), pp. 303–314.

McDavid, C. (2010). Public archaeology, activism, and racism: Rethinking the 'heritage' product. In: M.J. Stottman, ed., *Archaeologists as activists: Can archaeologists change the world?* Tuscaloosa: University of Alabama Press, pp. 36–47.

McGrath, A. (2015). Deep histories in time or crossing the great divide? In: A. McGrath and M.A. Jebb, ed., *Long history, deep time: Deepening histories of place*. Canberra: ANU Press, pp. 1–31.

McSorley, K. (2012). Introduction. In: K. McSorley, ed., *War and the body: Militarisation, practice and experience*. London: Routledge, pp. 1–39.

Meiklejohn, A. (1952). History of lung diseases of coal miners in Great Britain. III. 1920–1952. *British Journal of Industrial Medicine, 9*(3), pp. 208–220. doi:10.1136/oem.9.3.208

Merleau-Ponty, M. (2008). *The phenomenology of perception.* London: Routledge.

Meskell, L. (2002). Negative heritage and past mastering in archaeology. *Anthropological Quarterly, 75*(3), pp. 557–574.

Meskell, L. (2003). Memory's materiality: Ancestral presence, commemorative practice and disjunctive locales. In: R. van Dyke and S. Alcock, eds., *Archaeologies of memory.* London: Blackwell, pp. 34–55.

Meusburger, P., Heffernan, M. and Wunder, E. (eds.) (2011). *Cultural memories: The geographical point of view.* New York: Springer.

Meyers, M. and Woodthorpe, K. (2008). The material presence of absence: A dialogue between museums and cemeteries. *Sociological Review Online, 13*(5).

Mickel, A. and Knodell, A. (2015). We wanted to take real information: Public engagement and regional survey at Petra, Jordan. *World Archaeology, 47*(2), pp. 239–260.

Miller, D. and Parrot, F. (2009). Loss and material culture in South London. *Journal of the Royal Anthropological Institute, 15*(3), pp. 502–519.

Misztal, B. (2003). *Theories of social remembering.* Maidenhead: Open University Press.

Mixter, D.W. and Henry, E.R. (2017). Introduction to webs of memory, frames of power: Collective remembering in the archaeological record. *Journal of Archaeological Method and Theory,* 24(1), pp. 1–9.

Mookherjee, N. (2015). The raped woman as a horrific sublime and the Bangladesh war of 1971. *Journal of Material Culture,* 20(4), pp: 379–395.

Morgan, P. (2009). 'I was there too'. Memories of victimhood in wartime Italy. *Modern Italy,* 14(2), pp. 217–231.

Mulcahy, D. (2017). The salience of liminal spaces of learning: Assembling affects, bodies and objects. *Geografica Helvetica,* 72, pp. 109–118.

Mulhearn, D. (2008). Joint accounts. *Museums Journal,* 108(9), pp. 22–25.

Muzaini, H. (2015). On the matter of forgetting and 'memory returns'. *Transactions of the Institute of British Geographers,* 40, pp. 102–112.

Muzaini, H. (2016). Personal reflections on formal Second World War memorials in everyday spaces in Singapore. In: D. Drozdzewski, D. De Nardi and E. Waterton, eds., *Memory, place and identity: Commemoration and remembrance of war and conflict.* London: Routledge.

Navaro-Yashin, Y. (2009). Affective spaces, melancholic objects: Ruination and the production of anthropological knowledge. *Journal of the Royal Anthropological Institute* NS, 15, pp. 1–18.

Navaro-Yashin, Y. (2012). *The make-believe space: Affective geography in a postwar polity.* Durham, NC: Duke University Press.

Neri Serneri, S. (1995). A past to be thrown away? Politics and history in the Italian resistance. *Contemporary European History,* 4(3), pp. 367–381.

Newson, P. and Young, R. (eds.) (2017). Conflict: People, heritage and archaeology. In: P. Newson and R. Young, eds., *Post-conflict archaeology and cultural heritage.* London: Routledge, pp. 3–20.

Nora, P. (1984). *Les Lieux de Memoire.* Paris: Gallimard.

Nora, P. (1989). Between memory and history: Les lieux de memoire. *Representations,* pp. 7–24.

Nordstrom, C. (1997). *A different kind of war story (The ethnography of political violence).* Philadelphia: University of Pennsylvania Press.

Olick, J. (2008). From collective memory to the sociology of mnemonic practices and products. In: A. Erll and A. Nunning, eds., *Cultural memory studies: An international and interdisciplinary handbook*. Berlin: Walter de Gruyter, pp. 151–161.

Olivieri, L.M. (2017). Archaeology from below in Swat, Pakistan: Heritage and social mobilization in a post-conflict reality. In: P. Newson and R. Young, eds., *Post-conflict archaeology and cultural heritage*. London: Routledge, pp. 217–238.

Olstrom, E. (1996). Crossing the great divide: Coproduction, synergy, and development. *World Development*, 24(6), pp. 1073–1087.

Onciul, B., Stefano, M.L. and Hawke, S. (eds.) (2017). *Engaging heritage, engaging communities*. Woodbridge: Boydell Press.

Orange, H. (2010). Exploring sense of place: An ethnography of the Cornish mining world heritage site. In: J. Schofield and R. Szymanski, eds., *Local heritage, global context: Cultural perspectives on sense of place*. London: Ashgate, pp. 99–118.

Orange, H. (2015). *Reanimating industrial spaces: Conducting memory work in post-industrial societies*. Walnut Creek, CA: Left Coast Press.

Orange, H. (2017). Flaming smokestacks: Kojo Moe and night-time factory tourism in Japan. *Journal of Contemporary Archaeologym*, 4(1), pp. 59–72.

Orange, H. and Laviolette, P. (2010). Disgruntled tourist in king Arthur's court: Archaeology and identity at Tintagel, Cornwall. *Public Archaeology*, 9(2), pp. 85–107.

Orange, H. and Peters, C. (2011). The 'expert' amateur, professionalism and public engagement: The changing face of archaeology education in Cornwall from 1986 to 2011. *Cornish Archaeology*, (50), pp. 127–132.

Osborne, B.S. (2001). Landscapes, memory, monuments, and commemoration: Putting identity in its place. *Canadian Ethnic Studies*, 33(3), pp. 39–77.

Pain, R. and Staeheli, L. (2014). Introduction: Intimacy-geopolitics and violence. *Area*, 46(4), pp. 344–347.

Passerini, L. (1999). Resistances to memory, memories of resistance. In: H. Peitsch, C. Burdett and C. Gorrara, eds., *European memories of the Second World War: New perspectives on postwar literature*. Berghan: London, pp. 288–296.

Pavone, C. (1991 [2015 English edition]). *Una Guerra Civile. Saggio Storico sulla Moralità nella Resistenza*. Turin: Universale Bollati Boringheri.

Peitsch, H., Burdett, C. and Gorrara, C. (eds.) (1999). *European memories of the Second World War: New perspectives on postwar literature*. London: Berghan Books.

Perkins, C. (2004). Cartography – cultures of mapping: Power in practice. *Progress in Human Geography*, 28(3), pp. 381–391.

Petersen, R. (2005). Memory and cultural schema: Linking memory to political action. In: F. Cappelletto, ed., *Memory and World War Two: An ethnographic approach*. London: Berg, pp. 131–154.

Pezzino, P. (2005). The Italian resistance between history and memory. *Journal of Modern Italian Studies*, 10(4), pp. 396–412.

Philo, C. (2000). More words, more worlds: Reflections on the cultural turn and human geography. In: I. Cook, S. Naylor and R. Ryan, eds., *Cultural turns/geographical turns: Perspectives on cultural geography*. Harlow, Essex: Prentice-Hall, pp. 26–53.

Pickering, M. and Keightley, E. (2012). Communities of memory and the problem of transmission. *European Journal of Cultural Studies*, 16(1), pp. 115–131.

Pinder, D. (2005). Mapping worlds: Cartography and the politics of representation. In: A. Blunt, et al., eds., *Cultural geography in practice*. London: Arnold Publishers, pp. 172–190.

Pink, S. (2009). *Doing sensory ethnography*. London: Sage.

Portelli, A. (1997). *The battle of Valle Giulia: Oral history and the art of dialogue*. Madison: University of Wisconsin Press.

Portelli, A. (2003). The massacre at the Fosse Ardeatine: History, myth, ritual and symbol. In: K. Hodgkin and S. Radstone, eds., *Contested pasts: The politics of memory*. London: Berghan, pp. 29–41.

Price, P. (2010). Cultural geography and the stories we tell ourselves. *Cultural Geographies,* 17, pp. 203–210.

Puar, J. (2012) 'Becoming-Intersectional in assemblage theory.' philoSOPHIA, 2(1), pp. 49–66.

Radstone, S. (2005). Reconceiving binaries: The limits of memory. *History Workshop Journal,* 59, pp. 134–150.

Ram, K. and Houston, C. (2015). *Phenomenology in anthropology: A sense of perspective*. Bloomington: Indiana University Press.

Rednile Projects. (2018). Insider Art. Available at: www.rednile.org/public-realm/insider-art-kibblesworth/ [Accessed 21 Jul. 2019].

Renshaw, L. (2011). *Exhuming loss: Memory, materiality and mass graves of the Spanish civil war*. Walnut Creek, CA: Left Coast Press.

Richard, A. and Rudnyckyj, D. (2009). Economies of affect. *Journal of the Royal Anthropological Society,* 15(1), pp. 57–77.

Rico, T. (2008). Negative heritage: The place of conflict in world heritage. *Conservation and Management of Archaeological Sites,* 10(4), pp. 344–352.

Ricoeur, P. (1991). *A ricoeur reader: Reflection and imagination*. M. J. Valdés, ed. New York: Harvester Wheatsheaf.

Ricoeur, P. (1996). *The hermeneutics of action*. Edited by R. Kearney. London: Sage.

Ricoeur, P. (2004). *Memory, history, forgetting*. Translated by K. Blamey and D. Pellauer. Chicago: University of Chicago Press.

Riley, X. and Harvey, D. (2005). Landscape archaeology, heritage and the community in Devon: An oral history approach. *International Journal of Heritage Studies,* 11(4), pp. 269–288.

Rizvi, U.Z. (2006). Accounting for multiple desires: Decolonizing methodologies, archaeology, and the public interest. *India Review,* 5(3), pp. 394–416.

Rizvi, U.Z. (2018). Ambivalent fields: On the work of negative monuments. *American Anthropologist,* 14(3), pp. 542–543.

Robb, G. (1994). Environmental consequences of coal mine closure. *The Geographical Journal,* 160(1), pp. 33–40.

Rose, G. (1997). Situated knowledges: Positionality reflexivities and other tactics. *Progress in Human Geography,* 21, pp. 305–320.

Rouverol, A.J. (1999). "I was content and not content": Oral history and the collaborative process. *Oral History,* 28(2), pp. 66–78.

Said, E. (2000). Invention, memory and place. *Critical Inquiry,* 26(2), pp. 175–192.

Sather-Wagstaff, J. (2011). *Heritage that hurts: Tourists in the memoryscapes of 11 September*. Walnut Creek, CA: Left Coast Press.

Sather-Wagstaff, J. (2016). Making polysense of the world: Affect, memory, heritage. In: D. P. Tolia-Kelly, E. Waterton and S. Watson, eds., *Heritage, affect and emotion: Politics, practices and infrastructures*. London: Routledge, pp. 12–29.

Saunders, N.J. (2000). Bodies of metal, shells of memory: 'Trench Art' and the Great War Re-cycled. *Journal of Material Culture,* 5(1), pp. 43–67.

Schama, S. (2005). *Landscape and memory*. London: HarperCollins.

Schudson, M. (1995). Distortion in collective memory. In: D.I. Schacter, ed., *Memory distortion*. Cambridge, MA: Harvard University Press, pp. 346–363.

Seremetakis, N. (1993). The memory of the senses: Historical perception, commensal exchange and modernity. *Visual Anthropology Review,* 9(2), pp. 2–18.

Seremetakis, N. (1994). *The senses still: Perception and memory as material culture in modernity.* Chicago: University of Chicago Press.

Seremetakis, N. (2017). Performing intercultural translation. *Modern Greek Studies Yearbook,* 22/23, pp. 239–252. Minneapolis: University of Minnesota Press.

Shanks, M. (2012). *The archaeological imagination.* Abingdon: Routledge.

Silberman, N. (2014). Heritage interpretation and human rights: documenting diversity, expressing identity or establishing universal principles? In Ekern et al., eds., *World heritage management and human rights.* Abingdon: Routledge, pp. 33–44.

Simon, N. (2008). *The participatory museum.* San Francisco: Museum 2.0 Publisher.

Sium, A. and Ritskes, E. (2013). Speaking truth to power: Indigenous storytelling as an act of living resistance. *Decolonization: Indigeneity, Education and Society,* 2(1), pp. i–x.

Skeggs, B. (2004). Exchange, value and affect: Bourdieu and 'the self'. *The Editorial Board of the Sociological Review,* pp. 75–95.

Smith, L. (2006). *Uses of heritage.* Oxford: Routledge.

Smith, L. (2011). Affect and registers of engagement: Navigating emotional responses to dissonant heritage. In: L. Smith, G. Cubitt, R. Wilson and K. Fouseki, eds., *Representing enslavement and abolition in museums: Ambiguous engagements.* New York: Routledge, pp. 260–303.

Smith, L. and Campbell, G. (2011). Don't mourn, organize: Heritage, recognition and memory in Castleford, West Yorkshire. In: L. Smith, P. Shackel and G. Campbell, eds., *Heritage, labour and the working class.* London: Routledge, pp. 85–105.

Sobers, S. (2017). Ethiopian stories in an English landscape. In: H. Roued-Cunliffe and A. Copeland, eds., *Participatory heritage.* Facet Publishing. Available at: http://eprints. uwe.ac.uk/29163.

Solli, B. (1996). Narratives of Vøy: On the poetics and scientifics of archaeology. In: P. Graves-Brown, S. Jones and C. Gamble, eds., *Cultural identity and archaeology: The construction of European communities.* London: Routledge.

Stewart, K. (2007). *Ordinary affects.* Durham, NC: Duke University Press.

Stoller, P. (1997). *Sensual scholarship.* Philadelphia: University of Pennsylvania Press.

Strangleman, T. (2001). Networks, place and identities in post-industrial mining communities. *International Journal of Urban and Regional Research,* 25(2), pp. 253–267.

Stroulia, A. (2016). *Public archaeology: About the present and as a present.* Practicing Anthropology. Vol. 38, No. 2, pp. 32–36.

Suleiman, S.R. (2008). *Crises of memory and the Second World War.* Harvard: Harvard University Press.

Sultana, F. (2007). Reflexivity, positionality and participatory ethics: Negotiating fieldwork dilemmas in international research. *ACME,* 6(3), pp. 374–385.

Sumartojo, S. and Graves, M. (2018). Rust and dust: Materiality and the feel of memory at Camp des Milles. *Journal of Material Culture* 23(3), pp. 328–343. doi: 10.1177/1359183518769110.

Sumartojo, S. and Pink, S. (2018). *Atmospheres and the experiential world: Theory and methods.* London: Routledge.

Sumartojo, S. and Stevens, Q. (2016). Anzac atmospheres. In: D. Drozdzewski, S. De Nardi and E. Waterton, eds, *Memory, place and identity: Commemoration and remembrance of war and conflict.* London: Routledge, pp. 189–204.

Thomas, J. (1991). *Rethinking the Neolithic.* Cambridge: Cambridge University Press.

Thomas, J. and Ross, A. (2013). Mapping an archaeology of the present: Counter-mapping at the Gummingurru stone arrangement site, southeast Queensland, Australia. *Journal of Social Archaeology,* 13(2), pp. 220–241.

Thompson, P. (1998). *The voice of the past.* Oxford: Oxford University Press.

Thrift, N. (1997). The still point: Resistance, expressive embodiment and dance. In: S. Pile and M. Keith, eds., *Geographies of resistance.* London: Routledge, pp. 124–151.

Thrift, N. (2004). Intensities of feeling: Towards a spatial politics of affect. *Geografiska Annaler,* 86(B), pp. 57–78.

Tilden, F. (1977). *Interpreting our heritage.* Chapel Hill: University of North Carolina Press.

Till, K. (2005). *The new Berlin.* Minneapolis: University of Minnesota Press.

Tilley, C. (1994). *A phenomenology of landscape: Places, paths, and monuments.* Oxford: Berg.

Tilley, C. (2004). *The materiality of stone.* Oxford: Berg.

Tolia-Kelly, D.P. (2004a). Locating processes of identification: Studying the precipitates of re-memory through artefacts in the British Asian home. *Transactions of the Institute of British Geographers,* 29, pp. 314–329.

Tolia-Kelly, D.P. (2004b). Landscape, race and memory: Biographical mapping of the routes of British Asian landscape values. *Landscape Research,* 29, pp. 277–292.

Tolia-Kelly, D.P. (2006). Affect – an ethnocentric encounter? Exploring the 'universalist' imperative of emotional/affectual geographies. *Area,* 38(2), pp. 213–217.

Tolia-Kelly, D.P. (2010). *Landscape, race and memory: Material ecologies of citizenship.* London: Routledge.

Tolia-Kelly, D.P. and Crang, M. (2010). Affect, race, and identities. *Environment and Planning A: Economy and Space,* 42(10), pp. 2309–2314.

Tolia-Kelly, D.P., Waterton, E. and Watson, S. (2016). Introduction: Heritage, affect and emotion. In: D. Tolia-Kelly, E. Waterton and S. Watson, eds., *Heritage, affect and emotion: Politics, practices and infrastructures.* London: Routledge, pp. 1–11.

Trant, C. (1987). Art, landscape, and past. An artist's perspective. In: B.L. Molyneaux (1997), ed., *The cultural life of images: Visual representation in archaeology.* London: Routledge, pp. 11–21.

Trentmann, F. (2009). Materiality in the future of history: Things, practices and politics. *Journal of British Studies,* 48(2), pp. 283–307.

Trigg, D. (2012). *The memory of place: A phenomenology of the uncanny.* Athens, OH: Ohio University Press.

Tschuggnall, K. and Welzer, H. (2002). Rewriting memories: Family recollections of the national socialist past in Germany. *Culture and Psychology,* 8, pp. 130–145.

Tuan, Y. (2004). Sense of place: Its relationship to self and time. In: T. Mels, ed., *Reanimating places: A geography of rhythms.* Aldershot: Ashgate, pp. xx–09900.

Tuana, N. (2008). Viscous porosity: Witnessing Katrina. In S. Alaimo and S. Hekman, eds., *Material feminisms.* Bloomington: Indiana University Press, pp. 188–213.

Tudor, M. (2004). *Special Force: SOE and the Italian Resistance, 1943–1945.* Newtown: Emilia Publishing.

Turkle, S. (2007). *Evocative objects: Things we think with.* Cambridge, MA: MIT Press.

Turner, H. (2016). Critical histories of museum catalogues. *Museum Anthropology,* 39(2), pp. 102–110.

Turner, V. (1967). *The forest of symbols.* New York: Cornell University Press.

Tyner, J.A., Alvarez, G.B. and Colucci, A.R. (2012). Memory and the everyday landscape of violence in post-genocide Cambodia. *Social & Cultural Geography,* 13, pp. 853–871.

UNESCO. (1972). *General Conference 17th session*. Paris, 16 November 1972.

van Boeschoten, R. (2005). 'Little Moscow' and the Greek civil war: Memories of violence, local identities and cultural practices in a Greek mountain community. In F. Cappelletto, ed., *Memory and World War II: An ethnographic perspective*. London: Berg, 39–64.

Valentine, G. (1999). A corporeal geography of consumption. *Society and Space*, 17, pp. 329–351.

Vernant, J.P. (2001). *The universe, the gods, and mortals: Ancient Greek myths*. London: Profile Books.

Voss, B. (2008). Domesticating imperialism: Sexual politics and the archaeology of empire. *American Anthropologist*, 110(2), pp. 191–203.

Voss, B. (2015). What's new? Rethinking ethnogenesis in the archaeology of colonialism. *American Antiquity*, 80(4), pp. 655–670.

Voss, B. (2018). Archaeology is not enough: Witnessing the labor of Heritage stakeholders. *American Anthropologist*, 120(3), pp. 539–540.

Vuyk, K. (2010). The arts as an instrument? Notes on the controversy surrounding the value of art. *International Journal of Cultural Policy*, 16(2), pp. 173–183.

Walkerdine, V. (2009). Steel, identity, community: Regenerating identities in a South Wales town. In: M. Wetherell, ed., *Identity in the 21st century: New trends in changing times*. Basingstoke: Palgrave MacMillan.

Walkerdine, V. (2010). Communal belongingness and affect: An exploration of trauma in an ex-industrial society. *Body and Society*, 16(1), pp. 91–116.

Wall, S. (2006). An autoethnography on learning about autoethnography. *International Journal of Qualitative Methods*, 5(2). Available at: https://sites.ualberta.ca/~iiqm/back issues/5_2/PDF/wall.pdf

Waterton, E. (2005). Whose sense of place? Reconciling archaeological perspectives with community values: Cultural landscapes in England. *International Journal of Heritage Studies*, 11(4), pp. 309–326.

Waterton, E. (2014). A more-than-representational understanding of heritage: The 'past' and the politics of affect. *Geography Compass*, 8(11), pp. 823–833.

Waterton, E. (2015). Heritage and community engagement. In: T. Ireland and J. Schofield, eds., *The ethics of cultural heritage*. London: Springer, pp. 53–67.

Waterton, E. and Dittmer, J. (2014). The museum as assemblage: Bringing forth affect at the Australian War Memorial. *Museum Management and Curatorship*, 29(2), pp. 122–139.

Waterton, E. and Watson, S. (2015a). A war long forgotten: Feeling the past in an English country village. *Angelaki*, 20(3), pp. 89–103.

Waterton, E. and Watson, S. (2015b). Methods in motion: Affecting heritage research. In: B. Timm Knudsen and C. Stage, eds., *Affective methodologies: Developing cultural research strategies for the study of affect*. London: Palgrave MacMillan.

Watson, S. and Waterton, E. (2010). Reading the visual: Representation and narrative in the construction of heritage. *Material Culture Review*, 71, pp. 84–97.

Watson, S. and Waterton, E. (2011). Introduction: Heritage and community engagement. In: E. Waterton and S. Watson, eds., *Heritage and community engagement: Collaboration or contestation?* London: Routledge.

Watson, S. and Waterton, E. (2015). Methods in motion: Affecting heritage research. In: B.T. Knudsen and C. Stage, eds., *Affective methodologies: How to develop cultural research strategies for the study of affect?* Basingstoke: Palgrave Macmillan.

Wehner, K. and Sear, M. (2009). Engaging the material world; object knowledge and Australian Journeys. In: Dudley, S., ed., *Museum materialities: Objects, engagements, interpretations*. Routledge, pp. 143–161.

Wertsch, J. (2008). Blank spots in collective memory: A case study of Russia. *Annals of the American Academy of Political and Social Science,* 617(1), pp. 58–71.

Wetherell, M. (2012). *Affect and emotion: A new social science understanding.* London: SAGE.

White, H. (1990). Historical emplotment and the problem of truth. In: S. Friedlander, ed., *Probing the limits of representation: Nazism and the final solution.* Harvard: Harvard University Press, pp. 37–53.

Wiley, J. (2007). *Landscape.* London: Routledge.

Windle, J. (2008). The racialisation of African youth in Australia. *Social Identities: Journal for the Study of Race, Nation and Culture,* 14(5), pp. 553–566.

Witcher, R., Tolia-Kelly, D. and Hingley, R. (2010). Archaeologies of landscape: Excavating the materialities of Hadrian's Wall. *Journal of Material Culture,* 15(1), pp. 105–128.

Witcomb, A. (2010). Remembering the dead by affecting the living: The case of a miniature model of Treblinka. In: S.H. Dudley, ed., *Museum materialities: Objects, engagements, interpretations.* New York: Routledge, pp. 39–52.

Witcomb, A. (2012). On memory, affect and atonement: The long tan memorial cross(es). *Historic Environment,* 24(3), pp. 35–42.

Witcomb, A. (2015a). Cultural pedagogies in the museum: Walking, listening and feeling. In: M. Watkins, G. Noble and C. Driscoll, eds., *Cultural pedagogies and human contact.* London: Routledge.

Witcomb, A. (2015b). Toward a pedagogy of feeling: Understanding how museums create a space for cross-cultural encounters. In: A. Witcomb and K. Message, eds., *The international handbooks of museum studies.* Oxford: Wiley, pp. 321–344.

Witmore, C.L. (2007). Symmetrical archaeology: Excerpts from a manifesto. *World Archaeology,* 39(4), pp. 546–562.

Wolff, T. (2010). *The power of collaborative solutions: Six principles and effective tools for building healthy communities.* San Francisco, CA: Jossey-Bass.

Wood, D. (1992). *The power of maps.* New York: The Guilford Press.

Wood, D. (2002). The map as a kind of talk: Brian Harley and the confabulation of the inner and outer voice. *Visual Communication,* 1, pp. 139–162.

Wuming Foundation (2016). Available at: www.wumingfoundation.com/giap/2016/05/viaggio-nelle-nuove-foibe-3b-ritorno-dal-bus-de-la-lum-in-compagnia-della-xa-mas/ [Accessed 10 Mar. 2017].

Wylie, J. (2005). A single day's walking: Narrating self and landscape on the South West Coast path. *Transactions of the Institute of British Geographers,* 30, pp. 234–247.

Young, E. (1997). Towards a received history of the Holocaust. *History and Theory,* 36(4), pp. 21–43.

Zerubavel, E. (2003). *Time maps: Collective memory and the social shape of the past.* Chicago: University of Chicago Press.

Appendix

Appendix 1
Detailing a methodology

In this section of the book, I provide in detail the various stages in which the visualisation and mapping practices can come to life. In the following, I list step by step the ways in which the maps and visualisations in the previous chapters have been actualised. In the concluding section I propose ways to go beyond this static and largely paper-based technique to incorporate the wonderful challenges and affordances of open GIS, open source mapping and web publishing.

1 The design and planning of the maps: who, why, what

The first thing which the participants in the mapping or visualisation experiment is its purpose. Not only are we planning to make something happen, to make something stand out on paper or on the screen – we have to take into account the purpose and implications of any given visualisation exercise. In order to gauge these purposes and implications, we must turn to the community with whom we are conducting research and/or fieldwork. Only the community in question is able to figure out or to establish what they expect to gain or to learn from the process. In order to make sure that the community co-researchers and all participants gain from the experience, we might want to learn the extent of co-researchers' availability in the process. We should also ideally scope out the participants' emotional, political and practical investment and involvement in the idea of mapping memory/historical and archaeological sites/whatever they want to call it. Together with the other participants, we practitioners may find it useful to plan, and when not feasible, at least to anticipate who these artefacts are being produced for; and to identify the audiences and end users who the community expects to engage with the visualisations.

The decision to make these maps depends on a variety of factors, such as the intentions of the community and the timeline of research and fieldwork we intend to carry out. For instance, a community who fears misrepresentation in a heritage context might want to take direct action at the outset of the mapping strategy and establish spaces of contested/divergent memory and alternative perceptions at the centre of the mapping and fieldwork endeavour (see Hodgkin and Radstone, 2003). Starting or centring fieldwork and related activities such as ground photography, sound and video recording from selected foci of remembrance or contestation may then orientate the maps a certain way – as peaceful and creative tools for protest and assertion of local and culture-specific values. This scenario applied at Bus de La Lum, the anti-memory site which my co-researchers from the Cansiglio and Belluno areas of Italy explored (albeit reluctantly) in Chapter 5.

In some other cases, a community may wish to provide an overview of their own cultural and home milieu in order to preserve a holistic sense of place in the face of changing times or circumstances. The social group of map-makers in question may simply desire to sketch out their own perceptions and emotional attachments to place in order to protect and promote their own collective values for the world to appreciate and to open up their places and their memories to others. This was the case at Kibblesworth and Ryhope in northeast England, and the maps I presented in Chapter 6 serve as an opening up of positive, forward-looking place-making strategies in an area let down by socio-political neglect following deindustrialisation.

The desire to showcase complex, multi-layered and accretional local senses of place and place-specific values also applies to the north-eastern Italian communities who live and experience near and around the Monte Altare, in Chapter 3. In that chapter, we saw how local stakeholders, citizen experts and residents came together to care for the landscape and craft their everyday homescapes through caring and inclusive heritage and storytelling practices that centre on the iconic hill. The motivation for the latter map, which was part of a fieldwork endeavour of avocational archaeological societies in the Veneto region to reclaim their heritage sites as their own homescapes first and foremost, was not only political but also affectual. In other words, to become part of the story told about archaeological sites matters to the local communities, as the sites are part of the fabric of home.

The design of the maps and visualisations, then, foregrounds some aspects or provides an overview. Sometimes the mapping projects seek to do both. In Chapter 4, the community of Vittorio Veneto and environs decided to create two maps, one detailing wartime violence in the urban fabric of the town centre, and a wider one showing the movement of foreign bodies (German troops, Fascists militiamen from elsewhere in Italy). This double scale and double focus satisfied these specific groups of co-researchers that the problem of memory had been addressed from all necessary angles. The modes of design and planning could then be attributed to members and participants who had a specific interest in either focusing on the microscale of local vernacular violence, or those more invested in showing the broader mechanisms of invasion and occupation from further afield.

Whatever the scale or purpose behind the mapping exercises, the fundamental process tends to stay the same. A group of co-researchers flags the desire to condensate and open up memory in place. They come together and start planning the next steps. They try to decide, collectively what they want the maps to look like. The job of the academic or museum and heritage professional is then to create ways to enable multiple perspectives to blend together and appear side by side on the 'final' drafts, editing and advising on, but not appropriating, content.

A further fundamental issue to consider when doing this kind of methodology is the audience for the visualisations. Who is the exercise for? Who are the intended audience? Who is going to see, get access to, observe, interact with, engage with or use the maps? Will locals use the maps to guide outsiders in their discovery of a certain landscape, city, neighbourhood, block or even single building?

Logistics and portability also come into play. Where are these artefacts going to be stored? What modes of access and what affordances for use and tactile engagement will they offer? Are they going to go up as a poster in a classroom? Will they act as banners in a local museum or temporary exhibit? Will the maps become a permanent feature of some place? Will they appear in situ, next to or in the place they draw inspiration from and whose stories they tell? Will they appear remotely, detached from the original context of the things, people, stories and places they showcase?

Will the maps be static or portable? Are these items instead going to enjoy a dynamic materiality of use and reuse, of photocopying and portability as maps one can fold and take with them in a picket or rucksack? Will they guide people in the field, orientating bodies in the world using clues provided by the locals? Will archaeologists, tourists or students use them? Are they going to be reproduced, circulated and imitated? By whom? Are they an act of love or a pedagogical tool?

In the case of the group I worked with at Kibblesworth, the former residents of the Airey Houses were keen to make visible their lived experience as part of a community, part of the fabric of village life, in order to prevent the erasure of their affectual community of home – with its memories, places and materialities – from the re-built housing terrace on Beamish Museum's grounds.

The design phase takes into account the setup, size and format which any individual map is intended to be viewed at or used and circulated in printed form. A maximum of A2 print, with an optimal size of A3, is usually the preferred medium for portable printed maps. A2 was the chosen format by the Vittorio Veneto archival researchers in displaying the larger scale map of enemy action in the territory. A0 will probably be the format adopted by Beamish Museum for its Ryhope and Kibblesworth billboards and posters. The size of the artefact as it reaches a stage of completion and sharing which satisfies the community mapmakers will also determine the level and density of content and detail.

There are, inevitably, some commonalities in the design and implementation of a mapping plan. These commonalities are what determines the success or failure of a visualisation experiment, admittedly, but they also aid its potential replication by others (groups, schools, academics) if needed. In my experience, all visualisations develop along similar lines, spurred by similar kinds of agencies; most processes follow similar practices from inception to production. Usually, the initial phase of the crafting of these visualisations can consist of archival research or exploratory practices in the field. The reconnaissance of the terrain or the navigation of city streets can take place at the same time as desk-based learning and sharing of knowledge. I am a firm believer in the integration of multiple tasks and practices in the co-production process, whether in map making or writing up of reports. Everyone who is invested in a community-based project must be able to contribute something, from accessing parish records to taking photographs and uploading content on to the Web. The beauty of the process is the coming together and making together of something meaningful to local stakeholders and to outsiders or newcomers engaging with place.

2 The first phase: preparing for fieldwork

Fieldwork is common in all social sciences and participatory explorations are increasingly frequent across all manners of heritage and archaeological projects that foreground community and citizen expertise in the production of academic knowledge. In a project that involves community-led visualisations, memory maps and the like, the mapping activities shape the fieldwork design and the design and timing of fieldwork inevitably impacts on the extent and scope of the mapping. The chief benefit of the visualisation practices we are dealing with is that they can be developed around virtually any site, any period, any community anywhere in the world. It is a flexible, ductile and inclusive process that can give much satisfaction to a number of agents and stakeholders. The memory/heritage/knowledge visualisation method (and its product, the maps or visualisations themselves from draft to poster) is a form of making and learning on the ground, in the field, from the grassroots up to the academy through interlinked practices of visualisation, participatory processes, world-making and storytelling, which can then be used as channels for knowledge building in the community and beyond. Fieldwork and the visualisation process are inseparable, as more often than not, the maps we co-create with our community experts and residents reflect out shared experiences of being in the world – of being positioned some place.

An exploration of place which is inclusive, sensory-amplified and politically viable: this is the thrust behind most of the visualisation experiments I have taken part in. And most fieldwork agendas will be shaped by similar concerns. Look at the experiential heritage map of Col Castelir: we recorded the timings of ascent, the impressions we had while heading to the top and when we got there. Similarly, at the Caldevigo/Colle del Principe hill site, the community focused on their full immersion in the imaginative postmemory of Iron Age rituals. The senses took centre stage. The precise and accurate recording of the material culture from this site became secondary. And yet the participants were all involved in some way or other in the recovery and preservation of the objects from the Colle del Principe, and excluded from the site's formal publication – as had happened at the Monte Altare and the Col Castelir further northeast in the same region. Perhaps the Caldevigo community's wish to map themselves and their bodies in place first and foremost became a rebellion, a resistance. At the Col Castelir in the Cordignano area a similar resistance had taken place. The co-researchers from the local area also traced the progress of the participants' bodies in place, in a motion that went against the physical erasure of their experience from the archaeological discovery and paper-based, two-dimensional and largely quantitative engagements with the Col Castelir in mainstream academia. At any rate, the synergy between community's design and implementation of visualisation-based fieldwork takes several stages from preparation to participation and reflection. Later we consider the scale of the whole exercise from inception to printing and sharing.

Scale

What is the scale of the area we wish to visualise and 'cover'? The issue of scale was discussed in the chapters on the Italian civil war, where the community of Vittorio Veneto chose to actualise two maps, and this reflects the practice of a group who sought to provide as much context to the memory of the civil war as possible, without excluding the surrounding territory from the tragic stories of the town's violent past.

In deciding the scale of the field activities as well as the scale of the maps themselves, there are various factors to consider. The practicality of moving around in a large group can certainly be thrilling and create a real buzz, but also hinder reflective participation and prevent the systematic recording of impressions. If participants elect to take photographs or wish to record the positioning of their observation with an inexpensive hand-held GPS device, then large numbers of people and loads of chatter can distract from the accurate localisation of focal points that the community members wish to map. There is strength in numbers but also weakness. Therefore, the scale and details of the visualisations can determine group size. How many individuals are able to participate and over what length of time? How any field visits or trips do we need?

Then, there is the issue of print and dissemination and reuse of the material 'form' of the visualisations. Often communities wish for paper printed versions of the experiments to be generated so that they can be used and reused by the community and any site visitors who might wish to explore place using insiders' perception as a guide. The Beamish Museum maps, therefore, would become large scale billboards to accompany the recreated buildings in the 1950s town. Their scale matters.

When communities can make use of cheap or free territorial or city maps, they may wish to do so, as was the case for the Italian case studies thanks to the intervention of local councils. In Vittorio Veneto, the community chose aerial photographs from the 1950s and 1960s to map the wartime events and places that made up their shared geographies of home and grief. In the case of both maps, the small-scale city centre one and the territorial one, participants found that the terrain and actual topography of the chosen areas would serve better than abstract line drawn maps to reflect and refract the happenings and perceptions tied to this particular area.

In the case of the northeast of England and the memory maps that would end up at Beamish, we used modified versions of Google Maps, and in successive printed and annotated versions of local areas we used these as a canvas for the accumulation of memories, associations, infographics and photographs. A caveat is in order here. In some countries, such as Pakistan, the creation of new georeferenced cartographic data is prohibited; in these cases, communities and co-mappers might want to work on more imaginative, non georeferenced but still place-specific and perception-focused visualisations of their sense of identity and place.

Field presence

Most fieldwork will take place in the real world, as these visualisations tend to express an emplaced sensory engagement with place, a site, a town, or a memory, which are tethered somehow to actual locations. Yet this emplacement may take various forms and adopt various stages of presence or absence. On reflection, does the visualisation practice need any visits at all in order to be insightful and to satisfy the requirements and wishes of the community that is participating in the process? An absence of place can be as meaningful as its presence in the here and now. The more difficult or fraught the memories and associations with place, the more distance the map-makers might want to keep from the place itself. It bears to remember that at Bus de La Lum, the site of non-memory the community wished to map out was so infamously and tragically well known to everyone, that the physical act of visiting it was deemed unnecessary. The act of staying away from the sinkhole in fact produced a sense of catharsis from afar as far as I could see. The distance from the place we visualised became meaningful. The community members who took part had elected to take photographs in their own time, and some chose to find copyright-free images online rather than personally engaging with the site. Field presence, therefore, can mean different things to different communities and even to various subgroups within a larger number of community participants. The stakeholders and gatekeepers of the local Archaeoclubs insisted on being on site more times than the rest of the community map-makers at Caldevigo (fieldwork at the Colle del Principe), at the Col Castelir and at other sites whose maps do not appear in the present volume. An increased sense of ownership and investment in the archaeological past, as well as a more prominent engagement in the present, seem to dictate the extent to which various co-researchers elect or demand to be present on site during the proceedings.

In the following section, I move on to an overview of typical actions and processes that make up the dynamic 'data' collection that populates the maps and visualisations. The first phase is field based and interactional, meaning that all participants directly engage with the materials and/or places they wish to map out or visualise based on their preferences and abilities.

3 Walking, talking, snapping away: sensory histories on the go

Usually we start by field visits. By 'field' I mean literally any place that has a presence in the world – from a building to a city. The approach I usually take is to ask the group to be my guide, and to show me their places in their own terms. As local avocational community groups can be excluded from visibility in mainstream academia (in Italy this is more common than in other countries), I make it clear that their visibility and their presence are key. Therefore, the researcher or academic is dependent on what they are shown, where they are led, and whom they meet. The interaction with locals takes place in groups or individually, but the initiative should be led by local people, place or memory gatekeepers and any other stakeholders who are involved in the visualisation process.

At any given location, field participants and co-researchers lead the way, talking and explaining and telling stories. In some cases, local experts take the lead on archival searches and museum visits to take pictures of material which is in storage in archives or already displayed in museum cases, for instance. The sensory experience of discovering place and material culture in archives has been extensively covered in the literatures, so this is not the place to delve into its theoretical and practice implications. It is important however to stress the importance of sensory archive engagements in learning about heritage sites and things.

While some of the engagements that lead to mapping begin in the archive or store room, most experiments commence outdoors. Out there, where things are, participants feel that they can begin to delineate and shape a place's history and character. In the field, each participant expresses their own engagement differently. Some, or as I found, usually most of the participants take photographs that record their experience of being in place, and these images will end up on the aps (not all, but many do). The fleeting, ephemeral sense of togetherness and place-ness we experience in the field becomes part of the fabric of the maps through words, images and embodied encounters with the spatial entity or place we are mapping.

At Colle del Principe and Monte Altare, the act of ascending and navigating the hill sides of the former sacred sites in a still meaningful landscape became the core prism through which we made sense of the place. As mentioned earlier, the obfuscation of qualitative, anecdotal and personal accounts by non-professionals in mainstream publications authored by professional scholars can frustrate and offend local citizen experts. The centring of locally-made maps on the bodies and perceptive world of individuals remedies that top-down act of rejection.

Walking is the preferred method of navigation, with car travel as another option to cover larger distances. At Vittorio Veneto, walking around in groups led to a revisiting of the sites of wartime violence. Our moving around meant that we learned to recognise, to experience and to look at buildings that once hosted terrible secrets and now coexist with new uses and perceptions. We saw them as they are now, but remember or imagine what places used to be. We cannot hear the cries of pain or the silence of guerrilla actions, but we can imagine them pervading spaces long since changed.

At most sites, the experience of 'being there' became central on the map – often (if not always) the photographs taken by community participants usually become the visualisations' central mode of storytelling. The inclusion of visual material is paramount to the 'feel' of the maps for various reasons, but chiefly due to:

- Most sighted people are able to take photographs or create sketches and drawings or impressions of objects and places.
- The images lend an additional layer of sensory presence to the visualised locations, qualifying text or fixing material culture in place (as when the

Archaeoclub members located the surface finds on the map according to their discovery thereof).
- Images and photographs are concise media to convey information – spatial or otherwise – which take up less room on a map or composite visualisation than blocks of descriptive text.

Whatever the case may be, younger groups of map-makers and co-researchers always elect to have colour and texture on their own visualisations as, some have said, pictures bring you closer to a place. I can only agree with that sentiment!

4 Non-mobile memory work

The opportunity to move around in the world is not a given. In some groups, less mobile co-researchers have chosen not choose to navigate the sites even when, by their own admissions, the paths and roads where the mapping took place were reasonably accessible. As the key strength of community-led visualisation is their inclusivity and accessibility, to have non-mobile visualisations is a powerful way to open up the practice to non- or differently mobile residents and visitors. Whatever the reason, participants must feel at ease, in control of their own presence and contribution to the exercise. The contribution of non- or differently mobile co-researchers is priceless precisely in light of their own sensory perceptions of any limitations which place may present to differently able individuals. Non-mobile storytellers may elect to lead the way, metaphorically, from a position in which they themselves feel empowered.

Levels of personal mobility are not always the factor which leads to non-mobile engagements, however. In some other cases, as with the maps of Kibblesworth, focus groups replace the physical moving around in the world or on site, but focus reminiscence on specific sites and buildings to give more power and agency to selected sites at risk of being forgotten. So, at Kibblesworth, the main agency was a desire to keep the spirit of the Airey Homes alive, which involved an intimate reflection by a community of residents and friends. The reflection gave way to a sensory place-making practice which centred on the intimate spaces of buildings that no longer exist in their remembered/ experienced form. A site visit to their former street, Coltspool, would have achieved less of a potent reminiscence than our resident-specific interviews. Again, archive-based fieldwork can be an integral part of place engagement, and lead to astonishingly vivid representations and visualisations of a chosen entity in time and place. Whether moving or static, insights on place always centre on the individual and their perceptions, which is what empowers groups to express their sense of place.

5 First desk-based synthesis: going 'analogue' with paper and pens, or using software packages

Already at the inception phase of the visualisation, participants decide which 'base' the first map drafts and later versions will use as a context on which to focus impressions, stories, visual material and so on. Hence, earlier I have delineated a few options that communities and participants have in their decision on how to proceed. Once the map-makers elect what 'canvas' to adopt as a starting point – based on country and region-specific allowances for the creation of 'spatial' visual material – they are good to go and start adding things to the base map or image.

At some stage during fieldwork, participants start to think about the medium in which their visualisations will materialise. Do they want to go digital or do they want a nice large-scale print poster to display some place? Do they want to make postcards to distribute to site visitors or museum and library users?

In the days and weeks immediately after the first instalment of fieldwork (be it explorations, storytelling and/or archive based research and engagement), participants draft a first version of the map or maps they intend to create. This initial drafting can be accomplished in various ways to suit anyone's budget and time restraints.

In summation, there are two ways in which communities I have worked with and among have chosen to engage with the rough first drafts of their project:

* On paper, where an aerial photograph or an actual map is used participants choose to engage with the printed or physical medium by adding images, speech bubbles and other contextual data and information to the underlying 'canvas'. This is the option which schools usually tend to adopt. Teachers like having a large scale printout and work with post-it notes and coloured pens and pencils rather than on a screen, as direct engagement with the map and its components develops learners' visual and synthesis skills while also allowing for a hands-on feel for the project at this early stage
* Digitally, by scanning a base map or image and then adding elements of text and images to create a composite picture. The cheapest and most readily available way to achieve decent first drafts is by using Office's PowerPoint. Users can open a new slide, choose a background image to serve as the 'canvas' (map, drawing, photograph) that identifies the place they are visualising, and then add images and text boxes where desired in order to accomplish a collage-like first draft of the map. If participants elect to share the document using online archiving and sharing platforms (like Dropbox) or open source sites (Libre Office, and many more besides), they can take turns in editing the PowerPoint slide until they are happy to proceed and create more sophisticated, nuanced image files to reflect their work.

At later stages, participants can think about ways to develop more detailed and larger scale versions using different software, such as Adobe Photoshop and free online tools (Pixlr, Fotoflexer) and desk-based programmes like GIMP. 'GNU Image Manipulation Program' aka GIMP is an excellent free alternative to Adobe Photoshop which is easy to learn and use collaboratively. Pixlr is also an intuitive tool that allows users to create new images from scratch or to edit existing ones, such as making collages and the like. The possibilities are endless, even at this early stage of map design.

6 The collaborative editing phase: editing for content and design

This is the phase where the going gets tough for some, but interesting for most. The participants get together and elect which things, elements and items will populate the maps they are making. The competitive nature of some of the participants might get in the way of their better judgement on some occasions – especially when some participants have had the lion's share in the organisation of fieldwork or enjoy particular standing in their community.

The first thing to remember is that these kinds of visualisation are a platform for community members and stakeholders to create content and represent place in one way or another. The contents, design and 'feel' of the maps depend on the wishes and preference of the participants, and no one else. This way, the stakeholders can decide what appears and what does not appear on any given map or visualisation.

In most cases, there will be agreement on what is chosen to appear on any given version or draft of the map in production. Sometimes, however, there will be the inevitable divergence of opinion in regards to an item or choice of words to express a particular memory, emotion, impression and so on. The qualitative and ultimately subjective nature of these visual experiments is such that these incidents will occur, often. In my experience, they do not pose a barrier to the successful and often swift agreement over content. The thrill of competing views and versions of events, in some cases, is one of the most precious moments during the mapping fieldwork and production. It gets people talking, exchanging understandings and perceptions which may have hitherto been taken for granted or at face value. Statements are qualified – people ask questions and answer them (mostly). Imaginaries of place take centre stage along with material culture and more tangible elements of the perceptual or mnemonic landscape.

On a practical level, every participant wants something of theirs to appear on the map. Photographers will compete to have their snapshots 'come out' of the map to reflect the thick sensory experience of being there and snapping away. Storytellers will seek to get their words and memories on the map, to provide a more articulate and intimate account of places they know well. War veterans and their families and allies try to make their side of the story visible and relevant – they want to appear legitimate, acknowledged and they want to own their past. To make everyone's stories and images visible would be an impossible task, and this is why editing and decision-making become crucial to the whole endeavour. To create agreement over content is not to prioritise or order things in terms of worth or representativeness or value, it is to decide on the content and style based on what the community participants and future user can make of any given map. Most users appreciate a concise and not too crowded map – and if many visions and pictures would limit the navigation and comprehension of any of these visualisations, it is often a good idea to produce several maps incorporating various elements rather than cramming everything relevant and 'necessary' onto a single map.

7 Back in the field! Photography, blogging and more

When visualisation-making participants deem it necessary or desirable, they can go back out into the field or archive and acquire more experiences and or information at a later stage. The need to go back out and acquire more information does not mean that insufficient material was gathered the first time round – it means that the criteria for the making of the maps may have shifted. New stories emerge. New memories come to the surface, which the participants deem too important to leave off the map.

The participants need not have changed their minds – they may simply wish to incorporate more on something and less on something else. A participant's needs and requirements may have changed to a decision to talk about some aspects of their lives that they had previously chosen to be silent about. A former resident comes forward with more information about a dwelling, a neighbourhood, or a family or individual they have known. More photographs of places and things may be taken as new items and objects enter the story. Archival items lead to other archival items which make more sense when considered together. Participants may wish to disseminate their shared remembering practices by establishing a blog or a website at this stage too – but more about this later.

8 Working towards second drafts

The second draft is usually more satisfying to produce. When the participants are already fairly pleased with their first attempt, the scaling up (or down) of the visualisation to a smoother edit can be among the most empowering of experiences. When deciding on a second, more polished (often, but not always!) draft of their chosen maps, community participants can give free rein to their taste and capabilities – all according to an established consensus in content and design. At this stage of the production of the artefact/map, participants are usually pretty clear on what they want out of it and where the map or visualisation will ultimately end up – individuals concerned are focusing on its afterlife in terms of display, use and circulation. This is where the proper digitisation of the visualisations and their placing online or offline as independent artefact happens. Once the maps are satisfactorily 'complete' (but never final) the community map-makers and researchers have to decide how to make the best of it. Participants are proud of their effort, save for some highly sensitive contexts in which they have chosen to remain anonymous and disconnected from the materials being produced – as was the case in wartime memory work in Italy, for instance. On the other hand, if people have spent time participating in the visualisation experiments, they will want them to be seen and engaged with by others. The Beamish communities were keen to see the large versions of the maps go up in the Museum.

9 Printing, distributing and publishing

To tell the stories of their communities out of their 'native' context. Italian heritage groups and Archaeoclub members were keen to print out the maps as teaching aids in local schools but also as companion maps to site visitors. The Italian co-researchers were not particularly interested in the online afterlife of the maps for reasons of publicity and possible controversy in handling sensitive wartime memories (in Vittorio Veneto and at Bus de la Lum) or to avoid flaring up conflict with academic archaeologists whose use and publication of site-related materials might be in a state of embargo. Italian academic politics are very sensitive to the ownership of material which is still unpublished and not yet in the public domain . . . community experts and avocational archaeology stakeholders, therefore, need to preserve the smooth running of concerted practices at the regional level by avoiding conflict whenever possible. By publishing and distributing the heritage maps locally, stakeholders avoid any unforeseen pitfalls which may ensue their going 'global' via the world wide web and related complaints by academics who are writing up on those sites.

All in all, it bears to remember that the maps and visualisations are a tool to empower and not hinder or create problems for any given community – any strategy that serves to make visible but not problematize the local efforts to 'channel' and express place creatively and imaginatively are welcome, but we as academics should ensure that no harm will come to any of our co-researchers. Their participation is priceless, and our joint work should represent a pleasant and thrilling experience after we have left. We want our community based experts to recognise that they have enjoyed working with us and among themselves to make something positive – we do not want them to be liable for any trouble however small. As a learning and a *making* experience, the maps should reflect this synergy in a positive way.

Index

Printed in the United States
by Baker & Taylor Publisher Services

Printed in the United States
by Baker & Taylor Publisher Services